Surveying

Surveying

Gopal Narayan

RANDOM PUBLICATIONS
NEW DELHI (INDIA)

Surveying

ISBN 978-93-5111-436-9

Published in 2014 in India by
RANDOM PUBLICATIONS
4376-A/4B, Gali Murari Lal, Ansari Road
New Delhi-110 002
Phone : +91-11-43580356, +91-11-23289044
e-mail: randomexports@gmail.com, sales@randompublications.com, info@randompublications.com

Reprinted 2025

Type Setting by: Friends Media, Delhi-110089
Printed at : Replika Press Pvt. Ltd.

Preface

Surveying method is used for measurements of river channel and lake configurations. Often, less accurate methods can be used for this work than for water-level, recording stations, although the techniques are common. An alternative definition, from the American Congress on Surveying and Mapping (ACSM), is the science and art of making all essential measurements to determine the relative position of points or physical and cultural details above, on, or beneath the surface of the Earth, and to depict them in a usable form, or to establish the position of points or details. Furthermore, as alluded to above, a particular type of surveying known as "land surveying" is the detailed study or inspection, as by gathering information through observations, measurements in the field, questionnaires, or research of legal instruments, and data analysis in the support of planning, designing, and establishing of property boundaries. It involves the re-establishment of cadastral surveys and land boundaries based on documents of record and historical evidence, as well as certifying surveys of subdivision plats or maps, registered land surveys, judicial surveys, and space delineation. Land surveying can include associated services such as mapping and related data accumulation, construction layout surveys, precision measurements of length, angle, elevation, area, and volume, as well as horizontal and vertical control surveys, and the analysis and utilization of land survey data.

Surveyors use various tools to do their work successfully and accurately, such as total stations, robotic total stations, GPS receivers, prisms, 3D scanners, radio communicators, handheld tablets, digital levels, and surveying software. Surveying has been an essential element in the development of the human environment since the beginning of recorded history. It is required in the planning and execution of nearly every form of construction. Its most familiar modern uses are in the fields of transport, building and construction, communications, mapping, and the definition of legal boundaries for land ownership. The basic principles of surveying have changed little over the ages, but the tools used by surveyors have evolved tremendously. Engineering, especially civil engineering, depends heavily on surveyors. Whenever there are roads, railways, reservoir, dams, pipeline transports retaining walls, bridges or residential areas to be built, surveyors are involved. They establish the boundaries of legal descriptions and the boundaries of various lines of political divisions. They also provide advice and data for geographical information systems (GIS), computer databases that contain data on land features and boundaries. Surveyors must have a thorough knowledge of algebra, basic calculus, geometry, and trigonometry. They must also know the laws that deal with surveys, property, and contracts.

The texts are arranged in a lucid form and written in colloquial English. All the essential aspects of this subject have been included. Hopefully, the present study will prove very useful for students and teachers.

I thank all members of my team who have helped in the preparation of the book. My special thanks go to "Random Publications" who have published the book.

—Gopal Narayan

Contents

1

Introduction

Surveying

Surveying or land surveying is the technique, profession, and science of accurately determining the terrestrial or three-dimensional position of points and the distances and angles between them, commonly practiced by licensed surveyors, and members of various building professions. These points are usually on the surface of the Earth, and they are often used to establish land maps and boundaries for ownership, locations (building corners, surface location of subsurface features) or other governmentally required or civil law purposes (property sales).

To accomplish their objective, surveyors use elements of mathematics (geometry and trigonometry), physics, engineering and law.

An alternative definition, from the American Congress on Surveying and Mapping (ACSM), is the science and art of making all essential measurements to determine the relative position of points or physical and cultural details above, on, or beneath the surface of the Earth, and to depict them in a usable form, or to establish the position of points or details.

Furthermore, as alluded to above, a particular type of surveying known as "land surveying" (also per ACSM) is the detailed study or inspection, as by gathering information through observations, measurements in the field, questionnaires, or research of legal instruments, and data analysis in the support of planning, designing, and establishing of property boundaries. It involves the re-establishment of cadastral surveys and land boundaries based on documents of record and historical evidence, as well as certifying surveys (as required by statute or local ordinance) of subdivision plats or maps, registered land surveys, judicial surveys, and space delineation. Land surveying can include associated services such as mapping and related data accumulation, construction layout surveys, precision measurements of length, angle, elevation, area, and volume, as well as horizontal and vertical control surveys, and the analysis and utilization of land survey data.

Figure: *US Navy Surveyor at work with a levelling instrument*

Surveyors use various tools to do their work successfully and accurately, such as total stations, robotic total stations, GPS receivers, prisms, 3D scanners, radio communicators, handheld tablets, digital levels, and surveying software.

Surveying has been an essential element in the development of the human environment since the beginning of recorded history (about 6,000 years ago). It is required in the planning and execution of nearly every form of construction. Its most familiar modern uses are in the fields of transport, building and construction, communications, mapping, and the definition of legal boundaries for land ownership.

History of Surveying

Surveying techniques have existed throughout much of recorded history. In ancient Egypt, when the Nile River overflowed its banks and washed out farm boundaries, boundaries were re-established by a rope stretcher, or surveyor, through the application of simple geometry. The nearly perfect squareness and north-south orientation of the Great Pyramid of Giza, built c. 2700 BC, affirm the Egyptians' command of surveying.

A brief history of surveying:

- The Egyptian land register (3000 BC).
- A recent reassessment of Stonehenge (c. 2500 BC) suggests that the monument was set out by prehistoric surveyors using peg and rope geometry.
- The Groma surveying instrument originated in Mesopotamia (early 1st millennium BC).
- Under the Romans, land surveyors were established as a profession, and they established the basic measurements under which the Roman Empire was divided, such as a tax register of conquered lands (300 AD).
- The rise of the Caliphate led to extensive surveying throughout the Arab Empire. Arabic surveyors invented a variety of specialized instruments for surveying, including:
  - Instruments for accurate levelling: A wooden board with a plumb line and two hooks, an equilateral triangle with a plumb line and two hooks, and a reed level.
  - A rotating alhidade, used for accurate alignment.
  - A surveying astrolabe, used for alignment, measuring angles, triangulation, finding the width of a river, and the distance between two points separated by an impassable obstruction.
- In England, The Domesday Book by William the Conqueror (1086)
  - Covered all England
  - Contained names of the land owners, area, land quality, and specific information of the area's content and inhabitants.
  - Did not include maps showing exact locations.

In the 18th century in Europe triangulation was used to build a hierarchy of networks to allow point positioning within a country. Highest in the hierarchy were triangulation networks. These were densified into networks of traverses (polygons), into which local mapping surveying measurements, usually with measuring tape, corner prism and the familiar red and white poles, are tied. For example, in the late 1780s, a team from the Ordnance Survey of Great Britain, originally under General William Roy began the Principal Triangulation of Britain using the specially built Ramsden theodolite. Large scale surveys are known as geodetic surveys.

- Continental Europe's cadastre was created in 1808
  - Founded by Napoleon I (Bonaparte)
  - Contained numbers of the parcels of land (or just land), land usage, names etc., and value of the land
  - 100 million parcels of land, triangle survey, measurable survey, map scale: 1:2500 and 1:1250
  - spread fast around Europe, but faced problems especially in Mediterranean countries, Balkan, and Eastern Europe due to cadastre upkeep costs and troubles.

A cadastre loses its value if register and maps are not constantly updated. Because of the fundamental value of land and real estate to the local and global economy, land surveying was one of the first professions to require Professional Licensure. In many jurisdictions, the land surveyors license was the first Professional Licensure issued by the state, province, or federal government.

Surveying Techniques

Historically, distances were measured using a variety of means, such as with chains having links of a known length, for instance a Gunter's chain, or measuring tapes made of steel or invar. To measure horizontal distances, these chains or tapes were pulled taut according to temperature, to reduce sagging and slack. Additionally, attempts to hold the measuring instrument level would be made. In instances of measuring up a slope, the surveyor might have to "break" (break chain) the measurement- use an increment less than the total length of the chain.

Historically, horizontal angles were measured using a compass, which would provide a magnetic bearing, from which deflections could be measured. This type of instrument was later improved, with more carefully scribed discs providing better angular resolution, as well as through mounting telescopes with reticles for more-precise sighting atop the disc. Additionally, levels and calibrated circles allowing measurement of vertical angles were added, along with verniers for measurement to a fraction of a degree—such as with a turn-of-the-century transit.

The simplest method for measuring height is with an altimeter – basically a barometer – using air pressure as an indication of height.

But surveying requires greater precision. A variety of means, such as precise levels (also known as differential levelling), have been developed to do this. With precise levelling, a series of measurements between two points are taken using an instrument and a measuring rod. Differentials in height between the measurements are added and subtracted in a series to derive the net difference in elevation between the two endpoints of the series. With the advent of the Global Positioning System (GPS), elevation can also be derived with sophisticated satellite receivers, but usually with somewhat less accuracy than with traditional precise levelling. However, the accuracies may be similar if the traditional levelling would have to be run over a long distance.

Triangulation is another method of horizontal location made almost obsolete by GPS. With the triangulation method, distances, elevations and directions between objects at great distance from one another can be determined. Since the early days of surveying, this was the primary method of determining accurate positions of objects for topographic maps of large areas. A surveyor first needs to know the horizontal distance between two of the objects. Then the height, distances and angular position of other objects can be derived, as long as they are visible from one of the original objects. High-accuracy transits or theodolites were used for this work, and angles between objects were measured repeatedly for increased accuracy.

Turning is a term used when referring to moving the level to take an elevation shot in a different location. When land surveying, there may be trees or other obstructions blocking the view from the level gun to the level rod. In order to "turn" the level gun, one must first take a shot on the rod from the current location and record the elevation. Keeping the level rod in exactly the same location and elevation, one may move the level gun to a different location where the level rod is still visible.

Record the new elevation seen from the new location of the level rod and use the difference in elevations to find the new elevation of the level gun. Turning is not only used when there are obstructions in the way, but also when drastically changing elevations. You can turn up or down in elevation but the gun must always be at a higher elevation than the base of the rod. A level rod can usually be raised up to 25 feet high, which enables the gun to be set much higher. However, if the gun is lower than the base of the rod, you will not be able to take a shot because the rod cannot be lowered beyond the ground elevation.

Surveying Equipment

Figure: *A German engineer surveying during the First World War, 1918*

As late as the 1990s, the basic tools used in planar surveying were a tape measure for determining shorter distances, a level to determine height or elevation differences, and a theodolite, set on a tripod, to measure angles (horizontal and vertical), combined with the process of triangulation. Starting from a position with known location and elevation, the distance and angles to the unknown point are measured.

A more modern instrument is a total station, which is a theodolite with an electronic distance measurement device (EDM). A total station can also be used for levelling when set to the horizontal plane. Since their introduction, total stations have made the technological shift from being optical-mechanical devices to being fully electronic.

Modern top-of-the-line total stations no longer require a reflector or prism (used to return the light pulses used for distancing) to return distance measurements, are fully robotic, and can even e-mail point data to the office computer and connect to satellite positioning systems, such as a Global Positioning System. Though Real Time Kinematic GPS systems have increased the speed and precision of surveying, they are still horizontally accurate to only about 20 mm and vertically accurate to about 30–40 mm.

Total stations are still used widely, along with other types of surveying instruments, however, because GPS systems do not work well in areas with dense tree cover or constructions. One-person robotic-guided total stations allow surveyors to gather precise measurements without extra workers to look through and turn the telescope or record data. A faster but expensive way to measure large areas (not details, and no obstacles) is with a helicopter, equipped with a laser scanner, combined with a GPS to determine the position and elevation of the helicopter. To increase precision, surveyors place beacons on the ground (about 20 km (12 mi) apart). This method reaches precisions between 5–40 cm (depending on flight height).

Types of Surveys and Applicability

- *ALTA/ACSM Land Title Survey*: a surveying standard jointly proposed by the American Land Title Association and the American Congress on Surveying and Mapping that incorporates elements of the boundary survey, mortgage survey, and topographic survey.
- *Archaeological survey*: used to accurately assess the relationship of archaeological sites in a landscape or to accurately record finds on an archaeological site.
- *As-built survey*: a survey carried out during or immediately after a construction project for record, completion evaluation and payment purposes. An as-built survey also known as a 'works as executed survey' documents the location of the recently constructed elements that are subject to completion evaluation. As built surveys are typically presented in red or redline and overlayed over existing design plans for direct comparison with design information.
- *Bathymetric survey:* a survey carried out to map the topography and features of the bed of an ocean, lake, river or other body of water.
- *Boundary survey*: a survey that establishes boundaries of a parcel using its legal description, which typically involves the setting or restoration of monuments or markers at the corners or along the lines of the parcel, often in the form of ironrods, pipes, or concrete monuments in the ground, or nails set in concrete or asphalt.
- *Deformation survey:* a survey to determine if a structure or object is changing shape or moving. The three-dimensional positions of specific points on an object are determined, a period of time is allowed to pass, these positions are then re-measured and calculated, and a comparison between the two sets of positions is made.

- *Engineering surveys*: those surveys associated with the engineering design (topographic, layout and as-built) often requiring geodetic computations beyond normal civil engineering practice.
- *Foundation survey*: a survey done to collect the positional data on a foundation that has been poured and is cured. This is done to ensure that the foundation was constructed in the location, and at the elevation, authorized in the *plot plan, site plan*, or *subdivision plan*.
- *Geological survey*: generic term for a survey conducted for the purpose of recording the geologically significant features of the area under investigation. .
- *Hydrographic survey*: a survey conducted with the purpose of mapping the coastline and seabed for navigation, engineering, or resource management purposes.
- *Measured survey* : a building survey to produce plans of the building. such a survey may be conducted before renovation works, for commercial purpose, or at end of the construction process "as built survey"
- *Mortgage survey or physical survey*: a simple survey that delineates land boundaries and building locations. In many places a mortgage survey is required by lending institutions as a precondition for a mortgage loan.
- *Soil survey*, or soil mapping, is the process of determining the soil types or other properties of the soil cover over a landscape, and mapping them for others to understand and use.
- *Structural survey*: a detailed inspection to report upon the physical condition and structural stability of a building or other structure and to highlight any work needed to maintain it in good repair.
- *Tape survey*: this type of survey is the most basic and inexpensive type of land survey. Popular in the middle part of the 20th century, tape surveys while being accurate for distance lack substantially in their accuracy of measuring angle and bearing standards that are practiced by professional land surveyors.
- *Topographic survey*: a survey that measures the elevation of points on a particular piece of land, and presents them as contour lines on a plot.

Surveying as a Career

The basic principles of surveying have changed little over the ages, but the tools used by surveyors have evolved tremendously. Engineering, especially civil engineering, depends heavily on surveyors.

Whenever there are roads, railways, reservoir, dams, pipeline transports retaining walls, bridges or residential areas to be built, surveyors are involved. They establish the boundaries of legal descriptions and the boundaries of various lines of political divisions. They also provide advice and data for *geographical information systems* (GIS), computer databases that contain data on land features and boundaries. Surveyors must have a thorough knowledge of algebra, basic calculus, geometry, and trigonometry. They must also know the laws that deal with surveys, property, and contracts.

In addition, they must be able to use delicate instruments with accuracy and precision. In the United States, surveyors and civil engineers use units of feet wherein a survey foot is broken down into 10ths and 100ths. Many deed descriptions requiring distance calls are often expressed using these units (125.25 ft). On the subject of accuracy, surveyors are often held to a standard of one one-hundredth of a foot; about 1/8 inch. Calculation and mapping tolerances are much smaller wherein achieving near-perfect closures are desired. Though tolerances such as this will vary from project to project, in the field and day to day usage beyond a 100th of a foot is often impractical.

Licensing

In most of the United States, surveying is recognized as a distinct profession apart from engineering. Licensing requirements vary by state, but they generally have components of education, experience and examinations. In the past, experience gained through an apprenticeship, together with passing a series of state-administered examinations, was required to attain licensure. Now, most states insist upon basic qualification of a degree in surveying, plus experience and examination requirements.

The licensing process typically follows two phases. First, upon graduation, the candidate may be eligible to take the Fundamentals of Surveying (FS) exam, to be certified upon passing and meeting all other requirements as a surveying intern (SI), (formerly surveyor in training (SIT)). Upon being certified as an SI, the candidate then needs to gain additional experience to become eligible for the second phase. That typically consists of the Principles and Practice of Land

Surveying (PS) exam along with a state-specific examination. Licensed surveyors usually denote themselves with the letters P.L.S. (professional land surveyor), P.S. (professional surveyor), L.S. (land surveyor), R.L.S. (registered land surveyor), R.P.L.S. (Registered Professional Land Surveyor), or P.S.M. (professional surveyor and mapper) following their names, depending upon the dictates of their particular jurisdiction of registration.

In Canada, land Surveyors are registered to work in their respective province. The designation for a land surveyor breaks down by province, but follows the rule whereby the first letter indicates the province, followed by L.S. There is also a designation as a C.L.S. or Canada lands surveyor, who has the authority to work on Canada Lands, which include Indian Reserves, National Parks, the three territories and offshore lands.

In many Commonwealth countries, the term Chartered Land Surveyor is used for someone holding a professional license to conduct surveys. A licensed land surveyor is typically required to sign and seal all plans, the format of which is dictated by their state jurisdiction, which shows their name and registration number. In many states, when setting boundary corners land surveyors are also required to place survey monuments bearing their registration numbers, typically in the form of capped iron rods, concrete monuments, or nails with washers.

Building Surveying

Building surveying emerged in the 1970s as a profession in the United Kingdom by a group of technically minded general practice surveyors. Building surveying is a recognised profession in Britain, Ireland, Australia and Hong Kong. In Australia in particular, due to risk mitigation and limitation factors, the employment of surveyors at all levels of the construction industry is widespread. There are still many countries where it is not widely recognized as a profession.

Building Surveyors are trained to some extent in all aspects of property but with specific training in Building Pathology, as such they have a wide understanding of the end implications of decisions taken by more specific professions and trades during the realisation process, thus making them suitable for employment as Project and Property Managers on the client side (i.e. managing external contractors).

Services that building surveyors undertake are broad but can include:

- Construction design and building works
- Project management and monitoring
- Property Legislation advice
- Insurance assessment and claims assistance
- Defect investigation and maintenance advice
- Building surveys and measured surveys
- Handling planning applications
- Building inspection to ensure compliance with building regulations
- Pre-acquisition surveys
- Negotiating dilapidations claims

Building surveyors also advise on many aspects of construction including:

- design
- cost
- maintenance
- sustainability
- repair
- refurbishment
- restoration and preservation of buildings and monuments

Clients of a building surveyor can be the government agencies, businesses and individuals. Surveyors work closely with architects, planners, quantity surveyors, engineers, homeowners and tenants groups. A building surveyor may be called to act as an expert witness. It is usual for building surveyors to earn a university degree before undertaking structured training to become a member of a professional organisation.

With the enlargement of the European community, the profession of the building surveyor is becoming more widely known in other European states, particularly France, where many English-speaking people buy second homes.

Lidar Surveying – Three-dimensional laser scanning provides high definition surveying for architectural, as-built, and engineering surveys. Recent technological advances make it the most cost-effective and time-sensitive solution for providing the highest level of detail available for interior and exterior building work.

Land Surveyor

One of the primary roles of the land surveyor is to determine the boundary of real property on the ground. That boundary has already been established and described in legal documents and official plans and maps prepared by attorneys, engineers, and other land surveyors. The corners of the property will either have been monumented by a prior surveyor, or monumented by the surveyor hired to perform a survey of a new boundary which has been agreed upon by adjoining land owners.

Monuments are categorized into two groups which are known as natural and artificial. Natural monuments are things such as trees, large stones and other substantial, naturally occurring objects that were in place before the survey was made. An artificial monument is anything within the regulations that are usually placed at corner points by landowners, surveyors, engineers and others. They may be referred to as iron pins or pipes, stakes, trees, concrete monuments or whatever the surveyor decides to use at the time, within the regulations for the area. The courts have held that natural monuments control over artificial monuments because they are more certain in identification and less likely to be disturbed.

Over time, construction and maintenance of roads and many other acts of man, along with acts of nature such as earthquakes, movement of water, and tectonic shift can obliterate or damage the monumented locations of land boundaries. The land surveyor is often compelled to consider other evidence such as fence locations, wood lines, monuments on neighboring properties and recollections of people. This other evidence is known as Extrinsic Evidence and is a fairly common principle. Extrinsic evidence is defined as evidence outside the writings, in this case the deed. Extrinsic evidence is held to be synonymous with evidence from another source.

Today's land surveyor sets monumentation at actual physical points on the ground that define angle points of boundary lines that divide neighboring parcels. These monuments are most often 1/2" or 5/8" iron rebar rods or pipes placed at 18" minimum depth, but varies state by state. The more recent rods or pipes may have an affixed plastic cap over the top bearing the responsible surveyors' name and license number. Older monuments may exist such as old pipes, gun barrels, axles, mounds of stone, whiskey bottles, or even wooden stakes. In addition to rods and pipes, surveyors might use 4x4" concrete posts at corners of large parcels or anywhere that would require more stability (e.g. beach sand). They place them three feet deep. In places where there is asphalt or concrete, it is common to place nails or aluminium alloy caps to re-establish boundary corners. Marks are meant to be durable, stable, and as "permanent" as possible. The aim is to provide sufficient marks so some marks will remain for future re-establishment of boundaries. The material and marking used on monuments placed to mark boundary corners are often subject to state laws. Many states have laws that protect existing monuments and can have civil penalties if disturbed or destroyed.

Cadastral land surveyors are licensed by governments. In the United States, cadastral surveys are typically conducted by the federal government, specifically through the Cadastral Surveys branch of the Bureau of Land Management (BLM), formerly the General Land Office (GLO). They consult with USFS, Park Service, Corps of Engineers, BIA, Fish and Wildlife Service, Bureau of Reclamation, etc. In states that have been organized per the Public Land Survey System (PLSS), surveyors carry out BLM Cadastral Surveys in accordance with that system.

A common use of a survey is to determine a legal property boundary. The first stage in such a survey, known as a resurvey, is to obtain copies of the deed description and all other available documents from the owner. The deed description is that of the deed and not a tax statement or other incomplete document. The surveyor should then obtain copies of deed descriptions and maps of the adjoining properties, any records from the municipality or county, utility maps and any records of surveys. Depending on which region the survey is located in some or most of this information may not be available or even exist. Whether the information exists or not a thorough search should be conducted so that no records are neglected. Copies of deeds usually can be located in the county recorder's office and maps or plats can usually be found at the county recorder or surveyor's office.

These arrangements will vary state to state and survey system to survey system so some familiarity maybe needed. When all the records are assembled, the surveyor examines the documents for errors, such as closure errors. When a metes and bounds description is involved, the seniority of the deeds must be determined. The title abstract usually gives the order of seniority for the deeds related to the tract being surveyed and should be used if available. After this data is gathered and analyzed the field survey may commence. The initial survey operations should be concentrated on locating monuments. In urban regions or a city, monuments should be sought initially but in the absence of monuments property corners marked by iron pins, metal survey markers, iron pipes and other features that may establish a line of possession should be located.

When the approximate positions for the boundaries of the property have been located a traverse is run around the property. While the control traverse is being run, ties should be measured and all details relevant to the boundaries should be acquired. This includes but is not limited to locating the property corners, monuments, fences, hedge rows, walls, walks and all buildings on the lot. The Surveyor then takes this data collected and compares it to the records that were received. When a solution is reached the property corners that are chosen as those that best fit all the data are coordinated and ties by direction and distance are computed from the nearest traverse point. Once this has been established the features on the lot can be drawn, dimensions can be shown from these features to the boundary line and a map or plat is prepared for the client.

The Art of Surveying

Many properties have considerable problems with regards to improper bounding, miscalculations in past surveys, titles, easements, and wildlife crossings. Also many properties are created from multiple divisions of a larger piece over the course of years, and with every additional division the risk of miscalculation increases. The result can be abutting properties not coinciding with adjacent parcels, resulting in hiatuses (gaps) and overlaps. Many times a surveyor must solve a puzzle using pieces that do not exactly fit together. In these cases, the solution is based upon the surveyor's research and interpretation, along with established procedures for resolving discrepancies. This essentially is a process of continual error correction and update, where official recordation documents countermand the previous and sometime erroneous survey documents recorded by older monuments and older survey methods.

General

Basic field operations performed by a surveyor involve linear and angular measurements. Through application of mathematics (geometry and trigonometry) and spatial information knowledge, the surveyor converts these measurements to the horizontal and vertical relationships necessary to produce maps, plans of engineering projects, or Geographical Information System/ Land Information System (GIS/LIS). The highway surveyor must be adept at making the required measurements to the degree of accuracy required. Various types of engineering works require various tolerances in the precision of the measurements made and the accuracies achieved by these measurements.

The use of common sense and development of good surveying practice in all phases of a survey cannot be overemphasized. All conditions that may be encountered in the "real world" during the actual field survey cannot be covered in any manual. A manual may specify certain techniques, such as a certain number of repeated operations, to achieve a required accuracy.

The surveyor must then often use judgement based on the equipment being used and the field conditions encountered, to modify those techniques. Some field conditions (heat waves or wind for example) may make it impossible to perform some operations to a consistent degree of accuracy.

Accuracy

Accuracy is the degree of conformity with a standard or accepted value. Accuracy relates to the quality of the result. It is distinguished from precision that relates to the quality of the operation used to obtain the result. The standard used to determine accuracy can be:

A. An exact known value, such as the sum of the three interior angles of a plane triangle is 180°.

B. A value of a conventional unit as defined by a physical representation thereof, such as the international metre.

C. A survey or map value determined by superior methods and deemed sufficiently near the ideal or true value to be held constant for the control of dependent operations.

Although they are known to be not exact, higher order NGS control points are deemed of sufficient accuracy to be the control for all other less exact surveys.

Precision

Precision is the degree of refinement in the performance of an operation (procedures and instrumentation) or in the statement of a result. It is a measure of the uniformity or reproducibility of the result.

Accuracy versus Precision

The accuracy of a field survey depends directly upon the precision of the survey. Although through luck (compensating errors, for example) surveys with high order closures might be attained without high order precision, such accuracies are meaningless. Therefore, all measurements and results should be quoted in terms that are commensurate with the precision used to attain them. Similarly, all surveys must be performed with a precision that ensures that the desired accuracy is attained. However, surveys performed to a precision that excessively exceeds the requirements are costly and should be avoided.

Errors and Classification of Accuracy

General: Statistically speaking, field observations and the resulting measurement are never exact. Any observation can contain various types of errors. Often some of these errors are known and can be eliminated by applying appropriate corrections. However, even after all known errors are eliminated, a measurement will still be in error by some unknown value. To minimize the effect of errors, the surveyor has to use utmost care in making the observations and utilizing only calibrated equipment. However, a measurement is never exact, regardless of the precision of the observations.

Although this manual contains many guidelines and standards, the ultimate responsibility for providing surveys that meet desired accuracies remains with the field personnel. To fulfill this responsibility, the crew chief and his or her assistants must understand errors, including but not limited to:

A. The various sources of errors.

B. The effect of possible errors upon each observation, each measurement, and the entire survey.

C. Economical procedures which will eliminate or minimize errors and result in surveys of the desired accuracies.

Blunders: Many textbooks on surveying refer to a blunder as a gross error. One can easily make a case for a blunder to be considered an error. However, a blunder is really an unpredictable gross mistake made by the surveying team. It is not a hidden error that will go unnoticed, but usually it becomes apparent that something is wrong with the measurements. Examples of blunders are:

- Transposing two numbers (in field notes or computer input.)
- Misplacing decimal point.

- Incorrect reading (i.e. the foot value on a levelling rod.)
- Inadvertently altering set instrument constants in the middle of a project.
- Placing sighting device or the instrument at a wrong point.
- Misunderstanding verbal instructions or reading announcements (call out).
- Neglecting to level an instrument.
- Using the incorrect coordinates or benchmark values.

Blunders are caused by carelessness, misunderstanding, confusion, or poor judgement. They are, for the most part, avoided by alertness, common sense, and good judgement.

Blunders are detected and eliminated by using proper procedures, such as:

- Checking each recorded and calculated value.
- Making independent and redundant measure check observations and measurements.
- Making redundant measurements that allow closure computation of sections of the entire survey.

Small blunders are more difficult to detect and correct especially if the number of redundant measurements is too small. Therefore, surveys must be carried out with sufficient redundancy to prevent a blunder from going undetected. All blunders must be eliminated prior to correcting and adjusting a survey for errors.

Definition of Error

Error is the difference, after blunders have been eliminated, between a measured or calculated value of a quantity and the true or established value of that quantity.

Types of Errors

Excluding gross errors, which were discussed above, there are two general types of errors, systematic and random.

Systematic Errors

A systematic error is an error that will always have the same magnitude and the same algebraic sign under the same conditions.

In most cases, systematic errors are caused by physical and natural conditions that vary in accordance with known mathematical or physical laws. Systematic errors are caused by:

- Equipment out of calibration
- Use of insufficiently accurate computation equations (too few terms in a series.)
- Failure to apply necessary geometric reductions of measurements.
- Failure to apply necessary reductions of measurements due to weather related conditions.
- Personal biases of the observer.
- Use of incorrect units (feet instead of metres.)

A systematic error of a single kind is cumulative. However, several kinds of systematic errors occurring in any one measurement could compensate for each other. Some examples of systematic errors are:

- EDM that measures 99.95 feet while indicating a measurement of 100.00 feet.
- Refraction in vertical angles.
- Observer's tendency to sight on near or distant sights in a slightly different manner.

Although some systematic errors are difficult to detect, the surveyor must recognize the conditions that cause such errors. Once the conditions are known, the effect of these errors can be minimized as follows:

- Turning angles (with theodolite or total station) in direct and reverse modes.
- Balancing (maintaining similar distances between level and rod) foresights and backsights.
- Calibrating all surveying equipment.
- Calibrating EDM's yearly at a baseline calibration site.

When systematic errors cannot be eliminated by procedural changes, corrections are applied to the measurements. These corrections are documented in the user manuals of the equipment or in surveying textbooks. Undeterminable systematic errors can also be modelled into the adjustment computation, but surveyors should not rely on this. They must eliminate all the known systematic errors prior to proceeding with any adjustment of the survey data.

Random (Accidental) Errors

A random error (or accidental error) is an error produced by irregular causes that are beyond the control of the observer. They do not follow any established rule which can be used to compute the error for a given condition or circumstance of the observation. The occurrence, magnitude, and algebraic sign of a random error is truly random and cannot be predicted. For a single measurement, it is the error remaining in the measurement after all possible systematic and gross errors are eliminated. An important characteristic of the random error is that if we repeat the same measurement many times, the sum of all these errors tends to be zero. This is yet another good reason to make extra measurements beyond the required minimum.

An example of a random error is the personal reading error of any scale. An observer estimates the final reading that can be either high or low in estimation since exactness cannot occur. Unlike systematic errors, corrections for random errors cannot be computed directly. Random errors must be compensated by adjustments. The adjustment process computes adjusted observations for the actual ones in such a way that the remaining random errors are minimized. An example of such a process is computing an average distance from several measurements. The average represents the adjusted value for the distance for which the random error is minimized.

Random error obey the laws of chance or the random theory of statistics. Therefore, they are analyzed by applying the laws of probability. A complete discussion on the mathematical laws of probability is beyond the scope of this manual. The reference list at the beginning of this manual cites some excellent publications concerning the topic.

Sources of Error

Errors in measurements stem from three sources: personal, instrumental, and natural.

Personal Errors

Personal errors are caused by the physical limitations of the human senses of sight and touch. An example of a personal error is an error in the measured value of a horizontal angle, caused by the inability to hold a range pole perfectly in the direction of the plumb line. Personal errors can be either systematic or random. Personal systematic errors are caused by an observer tendency to react the same way under the same conditions. When there is no such tendency, the personal errors are considered to be random.

Common sense, self-calibration (estimating personal errors by experiments and experience) and attention to proper procedures generally keep such errors to a minimum.

Instrument Error

Instrumental errors are caused by imperfections in the design, construction, and adjustment of instruments and other equipment. Instruments can be calibrated to overcome these imperfections. Examples of instrument error are:

- Imperfect linear or angular scales.
- Instrument axes are not perfectly parallel or perpendicular to each other.
- Misalignment of various part of the instrument.
- Optical distortions causing "what you see is not exactly what you are supposed to see".

Most instrumental errors are eliminated by using proper procedures, such as observing angles in direct and reverse modes, balancing foresights and back sights and repeating measurements. Since not all instrument errors can be eliminated by procedures, instruments must be periodically checked, tested and adjusted (or calibrated.) Instruments must be on a maintenance schedule to prevent inaccurate measurements.

Natural Errors

Natural errors result from natural physical conditions such as atmospheric pressure, temperature, humidity, gravity, wind, and atmospheric refraction. Examples of natural errors are:

- A steel tape whose length varies with changes in temperature.
- Sun spots activity and its impact on the ionosphere, hence on GPS surveying.

Natural errors are mostly systematic and should be corrected or modelled in the adjustment. Some natural errors such as the effect of curvature and refraction can be eliminated by a procedure. The levelling procedure to eliminate curvature and refraction corrections is to average foresights and backsights.

Observation vs. Measurement

An observation is a single, unadjusted determination of a linear or angular value. A single reading of an angle or a single reading of an EDM is an observation. An observed value is a quantity that is obtained by instrumental measurement of the quantity. A direct observation is an observation of the desired quantity while an indirect observation is a quantity computed from direct observations. For example, rod readings in levelling are direct observations and the elevation difference between two points that is computed from these rod readings is an indirect observation.

A measurement is the entire process of obtaining a desired quantity. A measurement entails performing a physical operation that usually consists of several more elementary operations such as preparations (instrument calibration and setup), pointing, matching, and comparing (reading). The result of these physical operations renders a numerical value that is called a "measurement".

Surveys should be considered as measurements not as observations. With the advent of electronic readouts of linear or angular quantities everyone can make an observation. It requires a surveyor to make a measurement.

Linear Measurement

Taping

EDM instruments have largely replaced steel tapes in practically all measurements, except lower order work, such as close staking out, building measurements, etc. Wherever feasible, distances over 30 metres (100 feet) should be measured with an electronic measuring device. Accurate distances under 30 metres can be obtained with a calibrated steel tape.

Many surveyors believe that third order accuracy is a natural result of taping a distance. This is not true. Taped measurements will produce a linear accuracy of one part in 7,500 and yield a position closure of one part in 5,000 only if correct procedures are used. Such correct procedures would include standardization of tape, application of temperature correction, application of correct tension (particularly if tape is suspended), correct horizontal and vertical alignment of tape, and careful plumbing procedures. It is anticipated that taping will not be used in critical measurements and a detailed explanation of taping procedures will not be included in this discussion.

Electronic Distance Measuring

General Detailed operating instructions, instrument specifications and field adjustment information are included in the Operation Manual furnished with each instrument. The instrument manual should be kept with the instrument at all times.

Training Prior to performing field surveys, each operator should be thoroughly trained in the care and use of the measuring device and the allied equipment used therewith. The operator should be made fully aware of the instrument's limitations, possible causes of measurement errors, and have a thorough knowledge of the various functions performed by the instrument.

Checking Instrument When an instrument is received by a crew, whether new or transferred from another crew, the instrument should be checked on a known distance base line with the reflectors to be used with that instrument. All EDMs should be checked periodically, particularly prior to starting an important survey. NGS calibrated base lines are currently available in different locations within the state. Each calibrated base line has permanent monuments set to test instruments at several distances. A description of the test areas is included. Caution should be exercised to insure use of the most current data. Atmospheric Correction Atmospheric correction must be calculated and entered according to the instrument manufacturer's directions. Directions are usually supplied in the operator's manual for any instrument.

Common Sources of EDM Error

A. Setting Up The heavier EDM equipment puts an added strain on tripods and instrument stands. The tripods used to support EDM equipment should be sturdy and in good condition. Therefore, hinge and foot screws should be checked for tightness quite regularly.

B. Tribrachs Plumbing errors cannot be eliminated by measuring procedures. Therefore, tribrachs must be checked for adjustment (bubble and optical plummet) frequently. This includes not only the tribrachs used for the EDM instrument, but also those used with the reflectors. If a tribrach is accidentally bumped, dropped, or knocked over, it should be checked before any additional measurements are made.

C. Range Pole Mounted Prisms Range pole mounted prisms will seldom be used for measurements on the control traverse. When such prisms are used for tying in supplemental points and topographic features or staking out, an out of adjustment pole level can be the source of considerable error. Pole levels should be checked frequently and when in use should be attached securely to the pole in a position that can easily be viewed by the holder.

D. Reflections From Extraneous Objects Under most circumstances, an EDM measurement will be within the accuracy specified for that instrument even if the line of sight passes through leaves, fences, or other obstructions. However, such objects can sometimes reflect or interrupt the light rays and cause erroneous measurements. This occurs usually when the object is relatively close to the instrument. Roadside reflectors, windows, or other reflective objects in the path of or behind the prism can often cause erroneous measurements. When the line of sight cannot be cleared, such conditions should be recorded. Then, if poor closures result, those distances can be isolated and rechecked. When measuring various distances along a straight line, only one reflector should face the instrument. Otherwise, the instrument may be measuring to the wrong reflector.

E. Light Wave Skip All EDMs have the inherent capability of false readings due to light wave skips. The skip is generally in increments of 1, 10, 20 or more metres. Generally, the skip is of sufficient magnitude to alert the operator that an erroneous measurement is being made. However, at some distances the skip will be small and difficult to eliminate. Repeat measurements are often successful, but not always.

Quite often the small one metre skip will not be evident until the survey is closed. An analysis of the traverse can sometimes indicate the false measurement.

A routine method of checking for skip is by eccentric measurement. The reflector can be moved two or three feet on line and a check measurement taken. If the check measurement is near the eccentric difference, in all likelihood there is no skip present.

F. Improper Prisms or Preset Prism Constants Most EDM prisms have built in cross lens constants of 30 millimeters (0.098 feet). Each EDM manufacturer provides for their instruments' direct measurements by the combination of an internal adjustment within the instrument, and position of the prism in relation to the vertical axis of the prism.

In the final analysis, extreme caution should be exercised to make sure that the EDM prism constant setting coincides with that of the prisms being used. Be sure to check the EDM operator's manual and the specifications for the type prisms currently being used.

Number of Measurements

The number of measurements that should be made is a function of the characteristics of the survey. NGS has a set of standards and specifications for control surveys and the American Congress on Surveying and Mapping/ American Land Title Association (ACSM/ ALTA) has another set of specifications for land surveying work. Other organizations developed their own standards and specifications. These standards and specifications outline the number of linear measurements that are to be performed.

Under some circumstances, the distance between two points may be determined only on one line and only in one direction. In other words, reciprocal measurements or measurements to eccentric points may not necessary. A good rule of thumb to follow is that any time an angular direction is measured to a prism pole with a total station, a distance should be measured as well. There is very little extra effort involved in measuring the extra distance while the redundant measurement can provide a valuable check for the quality of the survey.

All base and control measurements should be the average of at least three measurements made in the standard (normal) measurement mode. Measurements made to set construction control stakes or points of equal importance should also be made in the standard mode.

Measurements made for topographic surveys, spot elevations, etc., can be made in the tracking mode. In order for the tracking mode to operate at the speed required, the instrument rounds off the displayed measurement.

Angular Measurement

Purpose

Points on the ground or on a map are related to each other through a horizontal distance and a horizontal angle (or direction.) Horizontal angular measurements are made between survey lines to determine the angle between the lines. A horizontal angle is the difference between two measured directions. Horizontal angles are measured on a plane perpendicular to the vertical axis (plumb line).

Vertical angular measurements are measured to determine slope of survey lines from the horizontal plane (level line). When the vertical angle is applied to the slope distance, the horizontal and vertical distances may be calculated. Vertical angles are measured on a plane passing through the vertical axis perpendicular to the horizontal plane. In order to facilitate the trigonometric calculations of horizontal and vertical distance, the reference or zero angle is on the vertical axis directly above the instrument, which is termed the zenith angle.

In the United States, the sexagesimal system of angular measurement is used. In the sexagesimal system, there are 360° in the circumference of a circle. The basic unit is the degree (°), which is further divided into 60 minutes (60'), and the minute is subdivided into 60 seconds (60"), and decimals thereof. Other angular unit systems utilize 400^g (grads) or 2p Radians per complete circle (360°).

Terms

The following terms are defined specifically for angular measurement. Their meanings may differ slightly in other contexts.

A. A pointing consists of a single sighting and circle reading on a single object.

B. An angular observation is a single, unadjusted determination of the size of an angle. A single angular observation is derived by subtracting the value of a pointing on a reference object from the value of a pointing on an observed station.

C. An angular measurement is the final determination of the magnitude of an angle before adjustment. Minimum angular measurement is the mean of at least two observations, one in the direct mode and the other in the reverse mode.

D. A reference object (RO) is a survey point that is used as an initial sight for orientation when measuring horizontal angles and "directions". The term, RO, will be used interchangeably with backsight (BS) in this manual.

E. A direction is the value of a clockwise angle between a backsight and any other survey point. The reference direction to the backsight can be either arbitrary or set to a desired value.

F. Setting a position is the act of setting a specified horizontal circle reading while the telescope is pointed toward a reference object. Generally, either zero degrees or the calculated "back azimuth" is used.

G. A direct reading is with the telescope in the upright (normal) position. An inverted or reverse reading is with the telescope inverted or plunged.

H. Turning a position is the act of making one direct and one reverse observation on each survey point to which a direction is required.

I. A repetition is a single observation (in a series of observations) of a horizontal angle, made with a repeating theodolite. This type of theodolite is rarely used nowadays. Information about repeating theodolites can be found in surveying textbooks.

J. Indirect measurement of an angle is a computed value of the angle from other data. For example angles of a triangle can be computed from distance measurements of its three sides. Orientation of the triangle is established by selected sides whose directions are known or measured.

Errors, Corrections, and Precautions

Direct measurement of angles and line direction by total station, theodolite, compass, or transit is familiar to all surveyors. However, many surveyors are not completely familiar with specific procedures that will achieve specified results. This section discusses errors involved in angular measurements and outlines procedures that will enable the surveyor to achieve specified results.

As mentioned earlier, errors in a measurement stem from various sources. Generally, angular measurements can be impacted by four classes of errors. These four classes are: instrumental, personal, natural, and miscellaneous errors. In the subsequent sections these factors and how to minimize them are discussed in detail.

Instrumental Factors

A. Adjustment Adjustments should be made at regular intervals and particularly before work on any control survey is started. Such adjustments should be made under the most ideal conditions available, normally in the highway yard or shop. Adjustment should be done in accordance with the user's manual of the specific instrument.

B. Servicing Instruments requiring major adjustments should be serviced at an authorized repair shop as specified by the Survey Operations Manager.

C. Level Bubbles and Optical Plummet Normal measuring procedures do not compensate for maladjustment of either the plate bubble(s) or the optical plummet. These components must be checked more frequently than others. On base and control traverse projects, the optical plummet should be checked at least once each day. The plate bubbles should be routinely checked on each setup.

D. Double Centering Double centering compensates for lack of adjustment of almost all components of the instrument and should be standard practice for all angles measured (or laid off with a transit). Double centering consists of two repetitions (one direct and one reverse) with a transit.

E. Parallax Parallax occurs when the focal point of the eyepiece does not coincide with the plane of the cross hairs. The condition varies for each observer because the focal length depends in part on the shape of the observer's eyeball. Parallax is also a major concern in the optical plummet.

1. When to Check Parallax should be checked by each instrument person when beginning to operate a new instrument or one that has been operated by someone else. The optical plummet should be checked on every setup, particularly if the instrument height is significantly different from the last setup.
2. How to Check Focus the telescope on some well defined distant object. Slowly move the head back and forth, about an inch from the eyepiece, while watching the relationship of the object to the cross hairs. If the object appears to move, parallax exists.
3. Eliminate Rotate the knurled eyepiece ring until apparent object movement is no longer present. It may be necessary to refocus the cross hairs.

Personal Factors

A. Setting up Instrument

1. Be sure the tripod is in good condition and all hardware is snugly fitted.
2. Push the tripod shoes firmly into the ground. Pressure should be parallel to each leg. Keep your foot lightly on foot piece when adjusting leg lengths.
3. Place the legs in a position that will require a minimum of walking around the setup. In windy conditions, additional stability can be achieved if one leg is set downwind.
4. If the ground is soft or muddy, drive long 50 x 100 millimeter (2" x 4") wedges or iron pipes 19 millimeters x 1 metre (3/4" x 36") in the ground to support the tripod legs. Use duck boards to support the instrument man.
5. On warm asphalt pavement set the tripod shoes on stakes that have been nailed to the pavement. Shading tripod feet from direct sun may also be helpful.

6. Be sure that the instrument is exactly over the point.
7. Check the optical plummet after the instrument is set up and just before moving to another point. If the instrument has moved, check the angle just measured.
8. Recheck the instrument level. The bubble should hold one position when the instrument is smoothly turned through one circle.
9. Protect the instrument from direct exposure to the sun. Use a parasol if necessary.

B. Setting Sights

1. When tribrach mounted targets are used, take the same precautions as when setting up an instrument. With this equipment, "forced centering" between targets and theodolite (and vice versa) will greatly decrease the effects of plumbing errors in traverse closures. Forced centering is especially beneficial in short course traverses. Forced centering is the traverse procedure whereby backsight, instrument point and foresight are "leapfrogged". Once a tribrach is set over a point, it must stay mounted on the tripod over that point for all uses. The instrument and sights are transferred from point to point without disturbing the tribrach setup.
2. Before picking up the instrument or the target, check to see that the tribrach or the sighting device has not moved.
3. When setting a pole sight, plumb it with a precision equal to that required for the total survey. A twenty second error results from a sight that is 0.1 metres (0.32 feet) out of plumb at 1,000 metres (3280.8 feet) away.
4. If a sight is set near ground level, check the line of sight for obstructions or for excessive heat waves. When excessive heat waves are present, ground level sights are not advisable.

C. Pointing

1. Tangent screw use When moving sighting device onto a target, always make the last turn of the tangent screw clockwise. Clockwise movement increases the tension on the loading springs. A final turn counterclockwise releases tension and the spring can hang up, causing a "backlash" error.

2. Cross Hair Use
   a. Consistency Sight each object with the same part of the cross hair, preferably near the centre of the field of view. This practice will minimize small residual adjustment errors.
   b. Technique The human eye can estimate the centre of a wide object more accurately than it can line up two objects. For this reason, different pointing techniques should be used depending on the type and apparent size of the sight in the telescope. When pointing on narrow sights, such as the centre of a red and white target or distant range pole, straddle the sight with the double cross hairs. When pointing on wide sights, such as a lath or range pole at close range, split the sight with the single cross hair.

D. Split Bubble Image A frequent error in zenith angle measurement is failure to adjust the split bubble image into coincidence before reading the angle. If coincidence is not made, the scale is not indexed to the vertical and significant errors can result. Always set the bubble before each reading.

E. Measuring Angles Measure angles as rapidly as comfortably possible with a uniform rhythm. Take the first reading at an object, rather than fidgeting with the tangent screw trying to improve the pointing. Too much pointing time increases the probability of error through instrument settlement or atmospheric changes. Speed should not be cultivated at the expense of good results. Accuracy is more important than speed.

F. Reading the Circles and the Micrometer
   1. Call Outs Carefully read and call out each reading to the recorder. Call out the entire reading each time so any large blunders will be caught. Have the recorder repeat the reading to the instrument man after it is recorded.
   2. When using the old style instruments, always check the ten minute interval. This is done by counting the number of graduations between opposite, corresponding degree marks, after the circle has been brought into coincidence.

G. Analyzing Observation Sheets Many recording errors and inconsistencies can be caught by carefully analyzing observation sheets. The following most important items should be checked in the field by the recorder and then rechecked in the office. They are:

1. Spreads between the seconds of direct and reverse readings should be consistent and in the same direction throughout the set.
2. Ten minute reading errors can frequently be located by examining a set of positions. This is the reason that it is highly recommended that the circle be closed (angle between foresight and backsight observed and recorded) for each observation. If one observation disagrees with all others by ten minutes, it is safe to assume that a reading or recording error was made.
3. When a direct and reverse observation of a position are in different minutes, be sure the average second value is coupled with the correct minutes value.
4. When checking the direction to a station and either the minutes or the seconds value of the backsight mean observation are greater than corresponding values to another station, be sure the subtraction is correct.
5. Using the entire set of averaged angles in the calculation will normally prevent errors in either of the above instances.

Natural Factors

A. Differential Temperatures Bright sunlight striking certain parts of the instrument may cause differential expansion of the metal components of the instrument, resulting in small errors. To minimize this error, it is recommended to work under a parasol.

B. Heat Waves Heat waves can cause distortion of lines of sight near a reflecting surface. The use of a tower can reduce the effect of heat waves. Working at night is another alternative available. Otherwise, unless it is urgent, on control surveys the work should be postponed until better conditions exist.

C. Phase If a sight is not evenly lighted on both sides, the instrument man will tend to point toward one side. This phenomenon, called "phase", can be reduced by using a target with a flat surface pointed directly toward the instrument. "Y" shaped targets are useful in reducing phase.

D. Diffraction If a line of sight passes very near a solid object such as a pole, light rays from a distant target may bend or diffract around the object, causing the object to appear in the wrong place. The closer the obstacle is to the instrument, the greater the diffraction. Diffraction can occur in either horizontal or vertical observations.

To detect diffraction, move the focusing knob slowly back and forth while watching the target. If the target appears to move relative to the obstacle and the cross hairs, the obstacle is causing diffraction. Offset the line to correct this condition.

E. Refraction When light waves pass from a medium of one density into a medium of a different density, the rays change in direction (bend). The change in direction is called refraction. Since sight lines are light rays, they are refracted, or bent, by changes in the atmosphere, causing small errors in angular measurement. Normally, the lateral refraction is insignificant in most surveys, but its effects can be further minimized by understanding and avoiding situations that generate the largest refraction of line of sight.

1. When the sun shines on a barren, dark surface, the surface warms relatively quickly. This warms the air and, if calm, it produces a column of warm, light air rising from the surface.
2. Other situations include:
   - Dark, freshly plowed fields lying between lighter coloured areas of growing crops.
   - Clear areas between heavy forests.
   - Large bodies of warm water between land areas.
   - Open valleys bordered by bluffs on either side. If a line must pass over a valley, set the observation points as far back from the edges of the valley as possible.
   - Air tends to layer parallel to the slopes of embankments or the base of foothills.
3. Minimizing Refraction When refraction is probable in angles to be measured, or is suspected in angles that have been measured, carefully examine the survey area and plan station locations to avoid problem conditions listed above.

Make your observations when a breeze keeps the air stirred and prevents layering, or on cloudy days or at night. Reobserve lines under different atmospheric conditions, preferably when the wind is from a different direction.

Miscellaneous Factors

A. Trigonometric Functions When trigonometric functions are used in computations, the precision of the computations is affected by the rates of change of the functions (the magnitude of the differences between the functions for a given angular increment). For example, to compute the impact of an angular error on the precision of the 'Sin' function at a particular angle, use the following equation:

$$precision\_of\_sin = \frac{\sin(\text{angle} + \text{error}) - \sin(\text{angle})}{\sin(\text{angle} + \text{error})}$$

B. Example: The precision of a sin of an angle of 5° with an error of 5" is 1:3610.

$$precision\,of\,\sin = \frac{\sin(5^o + 5") - \sin(5^o)}{\sin(5^o + 5")} = 0.000277\,or\,1:3610$$

C. For a 'Cosine' and 'Tangent' replace the 'sin' in the above equation with the respective trigonometric function. The following table shows the angles at which trigonometric functions have the highest and lowest precision.

	Sin	*Cos*	*Tan*
Highest precision	90°	0°	45°
Lowest precision	0°	90°	0° or 90°

D. The precision of the angular measurements might have to be increased to compensate for the large rates of change in the trigonometric functions.

E. This problem can be lessened by careful reconnaissance which:

1. Minimizes the use of such angles in the computations.
2. Provides sufficient checks that will ensure that desired accuracy is maintained (for example, cross ties in traverse).

F. Relationship Between Angular and Linear Measurement

1. Consistency When a survey involves both angular and linear measurements, maintain consistency (if practicable) between the precision of angular measurements and that of the linear measurements. In other words, keep the offsets in line, caused by errors in angular measurements,

approximately equal to the errors in linear measurements. For example, if distances are to be measured to a precision of 1/10,000, measure the angles to the nearest 20 seconds. (A 20 second error will result in an error of 0.10 metres (0.32 feet) offset in 1,000 metres (3280.8 feet).)

2. Angular errors and their corresponding linear errors per unit of distance can be computed from:

   Linear error = Tan(angular error) x *(distance of the line)*

Example: an angular error of 1" over a distance of 300metres (1000') is 1.5mm (0.005').

3. Inconsistency Often it is not practical to maintain the same precision in angular and linear measurements. In this case the accuracy of each measurement should be estimated and properly recorded. During the computation (or adjustment) of this data, each measurement should be weighted in accordance to its precision. Computation methods that do not accommodate for weighting measurements (such as the Compass Rule) should be avoided in this instance.

G. Curvature and Vertical Refraction Zenith angles measured for horizontal and vertical reduction of long lines require curvature and refraction corrections. The manufacturer's specifications for most total stations indicate that the earth's curvature and atmospheric refraction are internally computed and the corrected horizontal and vertical distances displayed. This correction can be eliminated (or balanced) by taking the mean of two reciprocal measurements. The reciprocal procedure means that a distance and a zenith angle are measured at point A to point B and then from point B back to point A. At longer distances (over 1000 metres (3280.8feet)), it is better to use this averaging solution than the on board correction factors provided within these total stations, even when they are in perfect adjustment.

Vertical Measurement

Purpose

In addition to the horizontal position of points and features as determined by the previously described linear and angular measurement, the complete survey requires vertical measurements. Such vertical measurements establish the elevation of points in relation to a datum that extends through and beyond the project limits. In the coordinate survey, the horizontal position is described by X and Y coordinates and the elevation described by the letter Z (or H.) The complete description of a point being (X,Y,Z).

Vertical angular measurements are also required to reduce slope measured distances to horizontal distances. Most EDM instruments used by the Department are equipped with zenith sensing devices that provide the reduction internally. These instruments display the operator's choice of slope, horizontal or vertical distance. Some older EDMs that are mounted on top of a theodolite measure the slope distance only. The theodolite is then used to make the angular measurement and the horizontal distance(X) is computed from the zenith angle(0) and slope distance(D). A correction for the earth's curvature and atmosphere must be applied to determine the true horizontal distance and difference in elevation (Y).

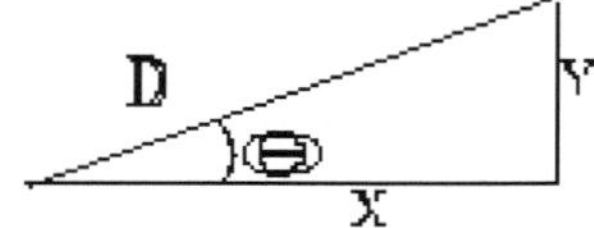

Importance

The determination of accurate elevations is an extremely important part of the information required for the design of highway projects. Grade lines, drainage structures and other highway features are designed in relation to existing and final elevations. Volumetric quantities are determined by preliminary (before) and final (as constructed) cross sections. Additionally, accurate elevations are very important in ensuring the reliability of photogrammetric mapping and orthophoto products.

Due to its importance in all other phases of the project development, vertical measurements to establish primary elevation control are made at an early stage of the survey. For example, in traversing, vertical angle elevations are established for all traverse control monuments to aid in the computation process. The control monuments become benchmarks. Subsequent fixed end loops between the control monuments established, again by differential levelling, are used to establish benchmarks for photogrammetric, preliminary, construction or other control networks.

Planning

By the time a project survey is completed from preliminary through the construction phase, each benchmark will have been used many times to provide the base for vertical measurement. Proper planning in anticipation of the future uses of vertical control benchmarks is as essential as that required for the horizontal control. Some considerations in that plan should be:

A. Location of the primary control (generally on the base traverse monuments).

B. Permanence (outside of anticipated construction limits), and type of monument set

- concrete monuments (permanent).
- Iron pins (semi permanent).
- wooden hubs (temporary)

C. Accessibility (on the Right of Way or other accessible lands).

D. Spacing (generally at 300 metres (1.000 feet) or less).

E. Visibility.

Methods

Vertical measurements are made directly or indirectly. The choice of the method and its procedures depend on present and future accuracy requirements and the relative cost. Considerations in selecting the method and procedures should include:

- Classification of controlling benchmarks. (The precision of the survey should be compatible with the accuracy of the controlling monuments.)
- Type of equipment available.
- Future survey needs.

Direct Vertical Measurement

This method means "the direct reading of elevations or vertical distances". The two common types of direct elevation determinations are readings from altimeters and from direct elevation rods. Tape (or Laser) measurement of a building height or depth of a mine shaft are also examples of direct measurements.

Elevations obtained by any of the above direct methods are less accurate and their application quite limited in highway surveys.

Indirect Vertical Measurement

Indirect vertical measurements require the use of calculations to determine elevations or vertical distances. Most vertical measurements made in highway surveys are made by indirect methods, such as spirit level (differential) and trigonometric levelling. There are numerous procedural systems used in both methods to achieve various levels of accuracy versus the expenditure of field and office time. In recent years, vertical measurements are made by GPS as well. It has been proven that careful GPS levelling can yield accurate elevations that are comparable with those of differential or trigonometric levelling.

Prior to the development of the total station, almost all vertical measurements on highway projects were made by differential levelling. By strict adherence to distance limitations and other procedural methods, trigonometric measurements can often be the best option for making vertical measurements.

Accuracy

Tolerances for the various types of control points and other elevation points are a function of the needs of a given project. The general requirements are cited here for reference.

A. Base traverse and monumented control points:

- Second order Class II misclosure (mm) is 8 x square root of D(D = shortest length of section in kilometers) or
- Maximum loop misclosure (mm) is 8 x square root of E (E = perimeter of loop in kilometers).

B. Secondary control, such as construction benchmarks, horizontal vertical photo control hubs, etc.:

- Third order misclosure (mm) is 12 x square root of D (D = shortest length of section in kilometers) or
- Maximum loop misclosure of 12 x square root of E (E = perimeter of loop in kilometers).

C. Vertical only spot elevations, slope stakes, etc. ± 0.015 metres (0.05 feet).

Differential Levelling

General

A. Equipment The standard instrument for all differential levelling is the pendulum type (automatic) level. The Department primarily uses Philadelphia two section, Lenker Rods, and fiberglass levelling rods. Each type of rod has its particular advantage under certain field conditions. Any rod used should be clean, "tight", and have properly indexed scales, Slip joint rods should be checked periodically for index.

B. Instrument Setups

1. Use a hand level in uneven terrain to aid in selecting setup and turning point sites.
2. Do not waste time by deeply imbedding tripod feet. Settlement is usually insignificant. Avoid setups on hot pavement or in spongy or muddy soil. If a setup in this area is unavoidable, be very careful around the instrument and constantly check level bubbles and backsights.
3. Set turning points so backsights and foresights are approximately equal. This compensates for curvature and refraction and for maladjustment of the instrument.
4. Use sight distances that best fit the terrain and are the most comfortable for the instrument person. Some operators can read the rod at much further distances than others. Sight distances should never exceed 75 metres (250 feet).
5. In steep terrain, place "turns" and instrument setups so they follow parallel paths (not along the same line).
6. Periodically test the level to be certain the pendulum compensator is working. Point on a "natural" sight with the telescope over a foot screw, and turn the screw back and forth, or lightly tap the instrument. If the cross hair dips and returns to its original position, the compensator is working properly.

Turning Points and Benchmarks

A. Establish benchmarks of the same physical quality as the technical quality of the levelling procedures. Set them in a stable, protected location.

B. Do not use spikes in utility poles. Do not use wooden stakes except as temporary benchmarks.

C. Make each turn stable and with a definite high point. If a turning point (TP) does not have a prominent point, mark the exact point with keel or paint. Depending on the soil, stakes driven at a slight angle make excellent TP's. A small piece of flagging will make them easily recoverable.

Rod Reading

A. Focus the eyepiece to eliminate parallax before any readings are made.

B. Do not deliberate over readings. Read and call them out in a moderate rhythm.

C. Turn through important points, rather than take "side shots". Benchmarks and photogrammetric control points should never be side shots.

D. Wherever possible, make it standard practice to plumb the rod with a rod bubble. In the absence of a rod bubble, the rod person should slowly sway the rod at and away from the instrument. The observer reads and records the lowest reading. The rod must be set on a sharp or rounded projection; otherwise the rod will rise as it is rocked and will result in a false reading.

E. Avoid low, ground-skimming shots where refraction might become pronounced. Also, avoid sighting close to obstructions that might diffract the line of sight; if possible, not closer than 0.3 metres (1 foot) to obstruction.

Single - Wire Levels

Single-wire levelling is the most common and widely used method of vertical differential measurement. With proper attention to procedural consistency, and by the use of several variable methods of elevation checks, third-order accuracy may be achieved with single-wire levelling. Some of the various methods will be described in general terms only. Detailed procedures may be found in the appropriate referenced publications at the beginning of this manual.

Computations - Normally, single wire notes are reduced to height of instrument (HI) and turning point (TP) elevations as the survey progresses. To check the elevations of benchmarks (BM's) that are turned through, differences in elevations, delta elevations, are also calculated. Basically, the difference of the sum of the plus rods (backsights) and minus rods (foresights) should equal the difference in elevation between the BM's. All side shots must be eliminated from the calculation. Adjustments Simple, single-wire level runs should be straight line adjusted. The closing error is prorated to each TP between two consecutive, controlling benchmarks.

$$C = E \cdot \frac{n}{N}$$

Where:

C – Correction applied at a TP.

E – Closure error of the levelling loop.

n – Number of turns to a given TP.

N – Total number of TP's in the loop.

Double Turning Point (TP) Levelling

This technique uses two parallel, independent foresight and backsight TP's for each HI. It is usually used for third order levelling or very special circumstances. Each pair of TP's is set, if possible, at an appreciable difference in elevation (preferably 0.3 metres or more). They are also set a few feet apart so the level will have to be rotated slightly between the two rod readings.

From each setup, single wire plus shots are read on both backsight TP's; minus shots are read on both foresight TP's. Notes are kept separately for each line of levels.

The adjusted elevations from the two lines of TP's are averaged.

The system has some advantages in that the HI is determined from each of the two lines and misreading or misleveling blunders can be isolated immediately.

The system is time consuming and both lines are run in the same direction, which may not cancel natural systematic errors.

Double Height of Instrument (HI) Levelling

This technique is similar to double TP levelling and is used for the same reasons as double TP levelling. A double line of levels is run through a single line of TP's. At each setup site, two HI's are established, at approximately 0.3 metres difference. From each HI the rod is read on the single backsight TP and on the single foresight TP.

The system has approximately the same advantages and disadvantages as for double TP levelling except that the difference in elevation can be immediately checked between the two TP's. If the difference is more than 0.001 metres (0.002feet), a third HI may be used.

Three Wire Levelling

When the distance between control benchmarks exceeds three kilometers, the system of three wire levelling can often be the most efficient method to establish project control benchmark elevations.

With this levelling technique, the cross hair and both stadia hairs are read to the nearest thousandth of a metre. (This system is widely used with an Invar rod and the readings taken to the nearest thousandth of a metre.) Stadia intercepts of plus and minus shots are accumulated. The running totals are constantly monitored so balance can be maintained between totals of foresight and backsight distances. The backsights and foresights should be balanced within 5 metres (20 feet) when setting BM elevations. This technique is generally preferred over the two previously described methods in that it is faster and, to a large degree, self checking. Since it is anticipated that this technique will be used most often in project control surveys, special requirements will be discussed in some detail.

A. Equipment

1. Instrument The Department generally uses pendulum levels with either a stadia ratio of 0.3 to 100 or 1.0 to 100. For precise levels, the 0.3 to 100 ratio is preferred because the stadia hairs are nearer the optical centre. It also permits a greater elevation difference between the level and the TP while keeping all three cross hairs on the rod.
2. Rods If Invar rods are not available, it is recommended that the best Philadelphia rod available be used on these surveys. Even new rods should be checked upon delivery, as some rods have been found to be not calibrated.

B. Procedures

1. Check the instrument and rod before starting each day's run, test the level for collimation error. Test at or near the first setup of the day and record the process in the field notes. If the error exceeds 0.005 in 200, the instrument should be adjusted. Any time the instrument has a severe jolt or bump, it should be readjusted. Check the rod in the raised position to ensure there is no index difference above and below the slip joint. The rod should be rechecked each time it is extended.
2. Setups Keep all sights within 75 metres (250 feet). When rejected readings average more than two in every ten, shorten the sighting distance.
3. Turning Points Railroad spikes, boat nails, wooden stakes or stakes may be used for TP's. If possible, all TP's should be left in place and flagged in case a complete or partial rerun is required. Permanent TP's should be numbered so they can be identified when recovered.
4. Benchmarks Establish all benchmarks prior to levelling. Check all found monuments that are to be incorporated in the level line for stability.
5. Rod Readings Plumb the rod with an accurate rod level. Start the rod reading with the top stadia wire, and progress

to the bottom wire. Estimate readings to 0.002 metres. Read at moderate speed without deliberations. Do not turn or pick up the instrument until the note keeper has verified the spread. If the spread between top and centre wire and bottom and centre wire exceeds 0.002 metres (0.0065 feet), reread all three wires. The original readings should be crossed out with a single line and new readings entered on the next line of the Field Book.

6. Reruns The highest order of accuracy required will normally be met by a single run of three wire levels. If the run fails to close within the tolerance specified, one of two problems probably exist.
   a. There may be a discrepancy between the elevations of the beginning and closing benchmarks. This should be the first possibility checked. Be sure that the government elevation data sheet is the last published for the two points. (There is a possibility that one or both of the benchmark elevations have been readjusted.) If the last information was used, carry the survey to the next benchmark.
   b. If the above does not account for the discrepancy, the run should be relevelled in the opposite direction. Relevelling at a different time of day is preferable.

Trigonometric Vertical Measurement

Trigonometric vertical measurement is a procedure whereby vertical differences in elevation are computed from slope distance and zenith angle measurements. The development and continuing refinement of total stations is making rapid changes in the use of trigonometric vertical measurements. These instruments are of varied vertical angle accuracy and certain procedural restrictions must be applied to equal the accuracy of differential levelling. Those restrictions required to meet the various tolerances will be discussed under each of the vertical traversing techniques described herein.

A. Applicability Vertical traversing is often the most practical (and economical) method for establishing elevations in rolling to steep terrain. It is useful for many types of surveys. Some of these are:
   1. Reconnaissance surveys.
   2. Control for aerial photography

3. Check levels for long differential lines.
4. Establishing elevations for datum adjustment.
5. Establishing low order benchmarks by precise vertical traversing. This would be done when accuracy is difficult to attain at reasonable expense by differential levelling.
6. Slope staking.
7. Cross sectioning.

B. Accuracy Attainable The accuracy that can be obtained from vertical traverses is sufficient for surveying work. The accuracy attained depends on the individual measurement accuracies of:

1. Slope distance With today's EDMS, slope distance accuracies are seldom significant in vertical measurement. For instance, with a zenith angle between 89 and 91 degrees, a slope distance error of one metre would result in a vertical error of less than 0.02 metre.
2. Zenith Angle The effect of zenith angle depends on the size of the angle and slope distance. To consistently attain good results, proper procedures must be used. Reciprocal observations minimize zenith angle and refraction errors. An erroneous zenith angle will result if the target and reflector are not properly spaced to provide parallel sight lines.
3. Height of Instrument (HI) and Height of the Reflector(HR) The major cause of error in trigonometric vertical measurement is the inaccurate determination of instrument and reflector heights. The net error that results from the two measurements produces a direct error in the difference in elevation. Aligning the height of the instrument and the reflector before making the measurements can reduce the effect of this error.

C. Calculations The calculation methods discussed herein provide satisfactory results.

The methods are described in principle only, as all such systems require consistent good practices in equipment operation as detailed in other sections of this manual. The equation for computing elevations using trigonometric levelling (without corrections for curvature and refraction) is:

$$H_B = H_A + S\,(\cos(Z)) + HI - HR$$

Where:

H_B - Elevation of point B (the new elevation to be established).

H_A - Elevation of point A (a point with known elevation on which the instrument is set up).

Z - Zenith angle from A to B.

S - Slope distance from A to B.

The term *S(cos(Z))* is the elevation difference component of the slope distance and it is denoted by DE.

There are three basic methods used in the computations to determine the elevation of a point from one or more points

1. Difference in elevation without HI or HR. This method is the most recommended one. It requires that the height of the rod and the height of the instrument are the same. As one can see from the above equation, if HI=HR then HI- HR=0. Thus, we do not need either of these heights. Most total stations have a special mark on the reflector as well as on the instrument that facilitates the alignment of these heights.
2. Difference in elevation with HR only. – When the instrument is set up at an arbitrary point (not a benchmark) and the reflector is held on a benchmark, HI can be computed adding HR and DE to the height of the benchmark. There is no instrument height measurement involved in this process. Once HI is established, additional points can be leveled ,even if HR is changed. The advantage of the above method is that the errors due to HI mismeasurement are eliminated. Once visualized, the systems are easy to use for remote wing points and spot elevations. They also allow the instrument to be set up at the best location for sight and distance.
3. Difference in elevation with HI and HR. This method requires the recording of the HI and HR for each observation. Calculations are done according to the above equation. While this is the least recommended procedure for trigonometric levelling, topographic and other constraints may dictate using this method.

Uses of Trigonometric Levelling

A. General – Elevation traversing, spot elevations, profiles of centerline and cross sections, drainage flow lines, or other required vertical information may be acquired by trigonometric levelling. Such elevations may be taken at identifiable points, such as staked centerline stations, or the horizontal position of said points may be determined during the same process.

B. Elevation Traversing – Trigonometric vertical measurements can be made at the same time that horizontal positions are surveyed or separately for vertical elevations only. Traverse point elevations should always be determined. Obtaining the elevations of traverse points require very little (if any) additional effort. The elevations of the traverse points can be used as a blunder check for more precise levelling of these points. They are also needed for reducing measurements to the state plane coordinate system,

C. Vertical Only Traverse On many occasions the elevation is required on points where the position is known (bridge ends, profile stations, etc.) or points that can be identified (wing points, etc.). Such elevations may be obtained by any of the three computation methods described previously. The accuracy required and position of the needed point relative to established horizontal and vertical control will generally determine the type of vertical traverse required.

D. Centerline Profile Centerline profiles are normally determined by differential levelling, but there are situations where trigonometric levelling may be the most economical. Much of the economy will depend on the location of control points in relation to the required profiles and the terrain involved. If the terrain is steep or rolling, requiring numerous "turns" with a level, trigonometric levelling may become the most economical method to obtain such elevations. Any of the same procedures as described for vertical measurement to remote points may be used. If all the spot elevations required are not visible from a nearby control point, the instrument should be set where one or more control points, and all the required spot elevations, are visible. In order to increase the reliability of the elevations, foresights and back sights should not exceed 300 metres (1,000 feet). All DE's should be read at least three times.

E. Roadway Cross Sections The standard method of obtaining field roadway cross section elevations is normally the most efficient. There are situations, however, where trigonometric levelling may be the most economical method. Such conditions may be where cross sections have to be extended beyond efficient taping distance, or where the terrain is steep and level turns are required. In such cases, the most efficient method of trigonometric levelling would be to set up over each cross section station turning right angles and measuring the distances and DE's with the EDM. This may be measured from the profile station, or by measuring the DE from a control point. If a uniform rod height can be used for the entire cross section, the relative elevation can be obtained by sighting another profile station, or a control point. If setting over the centerline station is impractical because of traffic, the same procedures may be used from an offset centerline.

F. Borrow Pit or Other Cross Sections Field cross sectioning or grid elevations of borrow pits, building sights, or odd shaped areas by trigonometric levelling can be fast and economical survey procedures. Several options are available using either trigonometric positioning and levelling, trigonometric positioning and differential levelling, or various combinations of the three techniques. The shape of the area and terrain involved will generally be the factors that would most influence which method would be the most economical.

1. Grids by Coordinates (Radial) The trigonometric positioning and levelling technique can be used from a single setup for which the coordinate position and elevation are known, or specifically established for the survey. Coordinates for each grid point can be prefigured and the azimuth and distance from the instrument tabulated for positioning the rod person at each grid point. Differences in elevations (DE's) would be read at each point and an accurate grid elevation computed. Total station tracking mode speeds up the operation. The primary advantage of this system is that very few stakes or flags are required and much of the control work can be prefigured at the office. If the grid area is quite large, additional setup points may be required.

2. Grid by Double Base Lines The trigonometric positioning and levelling technique is advantageous where the shape and topography of the area provides good visibility for the rod person. Two rows of grid lath are placed along one side and one end of the area. The rod person can "eyeball" grid points for either trigonometric or differential levelling on the unstaked points. This system requires some time to set up the control lath, but speeds up the location of grid points for the remainder of the survey.

3. Cross Sectioning from Base Line(s) Cross sectioning requires the establishment of one or more baselines. Normally, the instrument would be set up at each station and distances and DE's taken at each break along the cross sectional profile. Proper positioning of the base line(s) for good coverage and sights is the most critical part of this type of survey. This system is most advantageous where the terrain is rough and grids would not reflect the true topography.
4. Contouring Other possible uses of trigonometric levelling, such as contouring by random spot elevations, could also be considered for each survey. The point in all of the above being that trigonometric levelling gives the surveyor many options to do the work efficiently and effectively.

Chain (Unit)

A chain is a unit of length. It measures 66 feet, or 22 yards, or 100 links, or 4 rods (20.1168 m). There are 10 chains in a furlong, and 80 chains in one statute mile. An acre is the area of 10 square chains (that is, an area of one chain by one furlong). The chain has been used for several centuries in Britain and in some other countries influenced by British practice.

Origin

The chain was commonly used with the mile to indicate land distances and in particular in surveying land for legal and commercial purposes. In medieval times, local measures were commonly used, and many units were adopted that gave manageable units; for example the distance from London to York could be quoted in inches, but the resulting huge number would be unmemorable. The locally used units were often inconsistent from place to place.

In 1620, the clergyman Edmund Gunter developed a method of surveying land accurately with low technology equipment, using what became known as Gunter's chain; this was 66 feet long and from the practice of using his chain, the word transferred to the actual measured unit. His chain had 100 links, and the link is used as a subdivision of the chain as a unit of length.

In countries influenced by English practice, land plans prepared before about 1960 associated with the sale of land usually have lengths marked in chains and links, and the areas of land parcels are indicated in acres. A rectangle of land one furlong in length and one chain in width has an area of one acre. It is sometimes suggested that this was a medieval parcel of land capable of being worked by one man and supporting one family, but there is no documentary support for this assertion, and it would in any case have predated Gunter's work.

Contemporary Use

In Britain, the chain is no longer used for practical survey work. However it survives on the railways of the United Kingdom as a location identifier. When railways were designed the location of features such as bridges and stations was indicated by a cumulative longitudinal "mileage", using miles and chains, from a zero point at the origin or headquarters of the railway, or the originating junction of a new branch line. Since railways are entirely linear in topology, the "mileage" is sufficient to identify a place uniquely on any given route.

Thus a certain bridge may be said to be "at" 112m 63ch, meaning that it is at the location 112 miles and 63 chains (181.51 km) from the origin. In the case of the photograph the bridge is near Keynsham, that distance from Paddington station. The indication "MLN" after the mileage is the Engineers line reference describing the route as the Great Western Main Line, so that visiting engineers can uniquely describe the bridge they are inspecting, as there may be bridges at 112m 63ch on other routes.

The chain is not taught in British schools, but has survived for these reasons:

- Railways need to keep permanent records of as-built drawings of structures, and of the topography of routes and junctions;
- Chains and links are in many survey and real estate records;
- Miles and chains remain values familiar to many people.

Cricket Pitches

The chain also survives as the length of a cricket pitch, being the distance between the wickets.

Texas Chain

In Texas and elsewhere in the Southwestern United States, the vara chain of 20 varas (16.93 m, or ~55½ ft.) was used in surveying Spanish land grants.

Australian and New Zealand Use

In Australia and New Zealand, most building lots in the past were a quarter of an acre, measuring one chain by two and a half chains, and other lots would be multiples or fractions of a chain. The street frontages of many houses in these countries are one chain wide—roads were almost always 1 chain (20.1 m) wide in urban areas, sometimes 1.5 chains (30.2 m) or 2.5 chains (50.3 m). Laneways would be half a chain (10.1 m). In rural areas the roads were wider, up to 10 chains (201.2 m) where a stock route was required. 5 chains (100.6 m) roads were surveyed as major roads or highways between larger towns, 3 chains (60.4 m) roads between smaller localities, and 2 chains (40.2 m) roads were local roads in farming communities. Roads named Three Chain Road etc. persist until today. An acre is nominally the area within a rectangle 1 chain by 10 chains.

Chainage (running distance) is the distance along a curved or straight survey line from a fixed commencing point, similar to mileage.

North American Agriculture

In North America the chain is still used in agriculture: measuring wheels with a circumference of 0.1 chain (diameter H" 2.1 ft or 64 cm) are still common and readily available in the United States and Canada. For a rectangular tract, multiply the number of turns of one of these wheels for each of two adjacent sides, then divide by 1000 to get the area in acres.

Also in the United States the chain is normally used as the measure of the rate of spread of wildfires (chains per hour), both in the predictive National Fire Danger Rating Systems as well as in after-action reports. The term chain is used by wildland firefighters in day-to-day operations as a unit of distance. Under the U.S. Public Land Survey System, parcels of land are often described in terms of the section (640 acres or 259 hectares), quarter-section (160 acres or 64.7 hectares), and quarter-quarter-section (40 acres or 16.19 hectares). Respectively, these square divisions of land are approximately 80 chains (one mile or 1.6 km), 40 chains (half a mile or 800 m), and 20 chains (a quarter mile or 400 m) on a side. The use of chains and links is also commonly encountered in older metes and bounds legal descriptions.

The use of the chain was once very common in laying out townships and mapping the U.S. along the train routes in the 19th century. In the U.S. a federal law was passed in 1785 (the Public Land Survey Ordinance) that all official government surveys must be done with a Gunter's chain (also referred to as the "surveyor's chain"). Distances on township plats (US term for maps) made by the U.S. General Land Office are shown in chains.

Railroads in the United States have long since used decimal fractions of a mile, but the New York City Subway continues to use a chaining system using the 100 foot engineer's chain.

Ramsden's Chain

American surveyors sometimes used a longer chain of 100 feet (30.48 m), known as the *engineer's chain* or *Ramsden's chain.* The term chain in this case usually refers to the measuring instrument rather than a unit of length; the distances measured with such an instrument are normally measured in feet (and usually decimal fractions of a foot, not inches).

Other Instruments

Also in North America a modern variant of the chain as a tool is used in forestry for traverse surveys. This modern chain is a static cord (thin rope), 50 metres long, marked with a small tag at each metre, and also marked in the first metre every decimetre. When working in dense bush, a short axe or hatchet is commonly tied to the end of the chain, and thrown through the bush in the direction of the traverse, to ease working in dense forest.

Another version used extensively in forestry and surveying is the hip-chain. A hip-chain is a small box containing a string metre, worn on the hip. The user simply ties the spooled string off to a stake or tree and the metre counts distance as the user walks away in a straight line. These instruments are available in both feet and metres.

2

Linear Measurements

The aim of the inspection of a component is to see whether it lies within the prescribed limits or not and as to whether it meets the workmanship demanded by service specification. The critical examination of any component can be made by means of measuring instruments capable of dimensional control. Measuring instrument are designed either for line measurement (measuring distance between two lines, like steel rule), or end measurement (for measuring distance between two surfaces like micrometer screw gauge, etc). It is very difficult to use instruments designed for one system to be used for other system. The measuring instrument may be classified depending upon the accuracy that can be attained. The two categories are non-precision instrument and precision instrument.

Line Graduated Measuring Instruments

Line graduated Measuring instruments incorporate graduation spacing representing known distances. These are used for direct measurements of specific distances within their capacity range. The sensitivity of the measurement is dependent primarily on the instrument's basic design (the least distance between the individual graduations). The measuring accuracy is affected by the original accuracy of the graduation, the level of resolution of the graduation lines and of the read-out member, excessive thickness or poor definition of the graduation lines, the instrument's design and the general workmanship exercised in its manufacture, the geometric deficiencies resulting from flatness and parallelism errors or caused by deflection.

The precision of measurement made with line graduated instrument will also be a function of how truly are the actual distance to be measured associated with the corresponding instrument graduations. Observational errors may occur due to alignment deficiencies (due to improper coincidence of the distance boundaries with the selected scale graduations), parallax error when observing the scale and the object in a direction not perfectly normal to the surface being measured. However the observational errors can be reduced by means of auxiliary devices which improve the alignment of viewing and by keeping the measuring operation under control. But the instrument errors are inherent inaccuracies of the process and these must commensurate with the requirements of the particular measuring process.

Line graduated elements are used in great variety of measuring instruments. Differences exist in various essential respects, such as the intended use, the designed senility, the accuracy of execution, the level of sophistication with regard to read-out and many others.

Various types of line graduated measuring instruments are:

I- Line graduated rules and tapes ¯ These are used for direct length comparison and they have no auxiliary devices. These are available in widely different degree of accuracy to suit diverse requirements for plain length measuring tools.

II- Line graduated bar standards ¯ These are usually made in modified "H","U" or "X" cross-sectional forms, and the graduation lines are applied to that portion of the surface which lies in the neutral plane of the cross-section. It represents a very high level of accuracy.

Calipers

For the parts which can't be measured directly with the scale, the assistance of calipers is essential. Calipers thus act as accessories to scales. The caliper consists of two legs hinged at top, and the ends of legs span the part to be part to be inspected. This span is maintained and transferred to the scale.

It would be noted that calipers easily sense diameter (i.e. maximum distance) and transfer the distance between the faces to the rule in such a way as to reduce sighting errors and increase the reading accuracy.

Calipers can be either spring type or firm-joint type. Again under spring calipers we can have outside and inside calipers and under firm-joint calipers we have outside, inside, transfer and hermaphrodite calipers. In spring calipers; spring tension holds the caliper leg firmly against the adjusting nut. These are more accurate and permit accurate sense of touch in measuring.

Firm joint calipers work on the friction created at the junction of legs. These become loose after certain use. But they are easier to adjust and are particularly suitable for larger work. The various types of calipers could thus be classified as firm joint calipers and spring calipers.

Firm Joint Calipers

These are the devices for comparing measurements against known dimension. In the case of firm joint calipers, two legs (which are made from carbon and alloy steel containing not more than 0.05 per cent phosphorus and working ends suitably hardened and tempered to a hardness of 400 to 500 HV and measuring faces hardened and hardness of 650 ± 50 HV exactly identical in shape with the contact points equally distant from the fulcrum are joined together by a rivet. The legs are set correctly so that the working ends meet evenly and closely when brought together. In case the two legs are joined together by screw, nut and washer instead of rivet, then the threads of screw and nut should be full and true and the joints made such that these function freely with even tension without any undue play or stiffness. The component parts of the calipers should be free from seams, cracks, flaws and must have smooth bright finish.

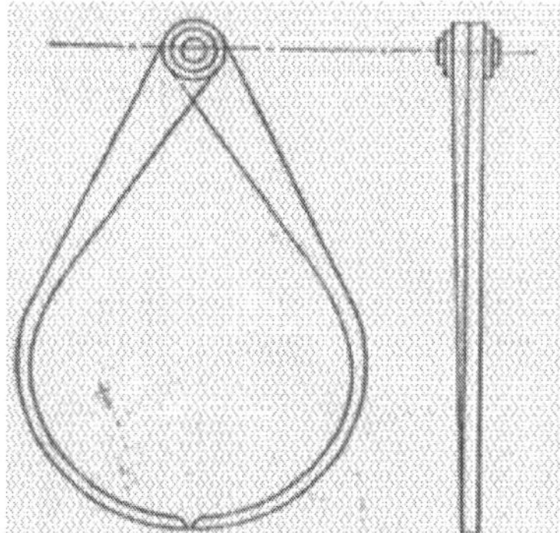

Figure: firm joint outside caliper

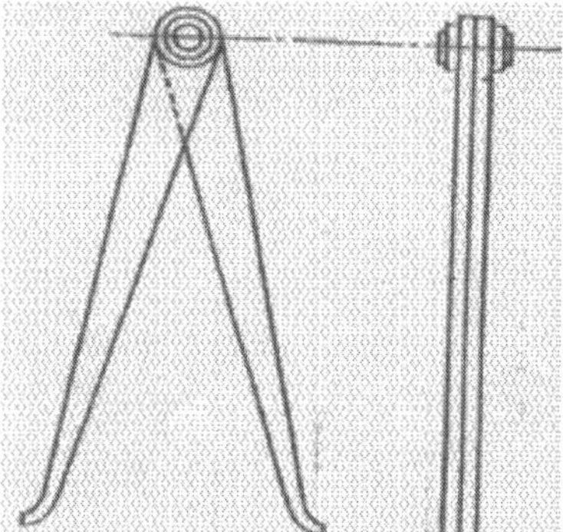

Figure: firm joint inside caliper

The firm joint calipers can be designed for inside measurement as well as outside measurements depending upon the shape of the legs. The distance between the roller centre and the extreme working end of one of the legs is known as nominal size and these calipers are available in the nominal size of 100, 150, 200 and 300 mm. The capacity of the caliper is the maximum dimension which can be measured by it. The capacity of the caliper should not be less than its nominal size. The legs these calipers are made of rectangular cross-section.

For assuring that the calipers function satisfactorily, it is essential to ensure that each caliper works smoothly and retains its size when adjusted to three test pieces of the sizes at extreme (lower and upper) ends and middle of the instrument range. The accurate use of calipers depends upon sense of feel of operator. For this purpose, caliper should be held gently, and held square to the work, and only light gauging pressure should be applied.

Firm Joint Divider

These are used for scribing arcs and circles. The distance between fulcrum roller centre and the extreme working end of one of legs is known as the nominal size and these are available in the size of 100, 200 and 300 mm. The working ends of the divider legs (up to 10 mm from extreme tips) are suitably heat treated in order to impart a hardness which will make it capable of scribing a clearly defined line across the face of a block (having hardness of 620 to 660 HV) and still the working ends should be sharp and visual examination should not reveal any sign of damage or failure. The working legs are made very sharp so that the trace on the test block does not show any sign of damage or failure. The fulcrum roller is hardened and tempered to a hardness of 625 HV and joint made such that the divider shall be capable of smoothly retaining its size. The rigidity of the leg and the proper functioning of the divider is judged by clamping securely one leg of the divider in a vice at a point about 13 mm from the fulcrum and opening the other leg slightly such that both the legs are in a horizontal plane. A test load of 4.5 kg is then gradually applied by suspension from the free leg as close to the working end as possible. On removal of the load, the joint should not be displaced and the legs should not show any signs any of permanent set.

Spring Calipers

The legs of spring calipers are made from suitable alloy steel, measuring faces properly heat-treated and hardened to hardness of 650 ± 50 HV. These should be free from seam; cracks, flaws etc. and all component parts should be smoothly finished. The spring which is made from carbon spring steel is properly hardened and tempered to hardness of 470 to 520 HV and all components are manufactured and assembled in such a way as to achieve and maintain the smooth adjustment and efficient operation.

The working ends of the legs of each spring caliper should be identical in shape and have contact points equally distant from the fulcrum. The measuring ends of each pair are so shaped that all sizes within its capacity may be readily transferred from measuring scales or gauge. The cross-section of the legs may be either of rectangular or circular shape. The calipers are adjusted to a set dimension by means of either a knurled solid nut or a knurled quick-action release nut operating on a finely threaded adjusting screw, the latter one being preferred. The top portions of the legs are located in a flanged fulcrum roller and held in position by a spring in order to maintain the alignment of the working ends. The spring provides sufficient tension to hold the legs rigid at all points of the adjustment. The friction between the adjusting nut and the leg is minimized by using a separate washer under the nut. One end of the adjusting screw is securely hinged to a leg of the caliper and a steel ball positively fixed to the free end of the screw for the purpose of retaining the adjusting nut. The inside and out side spring calipers are available in size of 75, 100, 150, 200, 250 and 300 mm. A good caliper should be able to retain its size when adjusted to three pieces near the lower and upper extremes of its range and in the middle.

According to Other Consideration, Calipers Could be Classified as:

(i) Outside calipers,

(ii) Inside calipers,

(iii) Transfer calipers,

(iv) Hermaphrodite calipers.

Outside Calipers

These are designed to measure outside dimensions. The accuracy in caliper measurement depends upon the inspector's sense of feel. When a caliper has to be set from the scale to control the dimension of any component, then leg of the caliper must be held firmly against the end of the proper dimension by adjusting nut with the thumb and forefinger.

When caliper is used to measure the outside diameter of a cylinder, then it must be held in hand with its nut between fore-finger and thumb, and one end of leg touching with the surface of cylinder. The other leg is moved to and from the opposite surface so that other leg just passes through the cylinder. The caliper reading is transferred to the scale by holding one leg of the caliper against the scale firmly and the reading taken at the other end of leg.

For accurate setting, the distance between the outside calipers may be set by slip gauge or by micrometer anvils. Sometimes it may be required to transfer this dimension to caliper.

Inside Calipers

While using inside calipers, the use of hook scale to measure the distance between its legs is desirable. With ordinary scale, one leg of inside caliper must be coincided with a vertical brace and second leg adjusted accordingly. It will be best if the legs of the inside caliper are adjusted with a micrometer.

Transfer Calipers

These are used for measuring recessed areas from which the legs of the calipers can't be removed directly but must be collapsed after the dimension has been measured. In the transfer calipers, an auxiliary arm is provided to preserve the original setting after the legs are collapsed.

Hermaphrodite Calipers

It is also known as odd leg caliper and is a scribing tool consisting of one divider and one caliper leg. Thus it is used in layout work. It can have two types of legs, viz., notched legs or curved leg. In order to avoid excessive wear of nut while closing the legs of spring calipers, the legs must be first collapsed. The caliper must be held square to the work when checking. Also when measuring the outside diameter, a caliper should not be forced upon the work. Forcing the caliper upon the work results in springing of legs and the false reading.

Dial Calipers

These are typical of all hand-held calipers. These provide a typical direct-reading capability of 0.02 mm. These units are quite flexible, offering a typical measurement range of 150 mm with accessories available for range extension, as well as specific measurement accessories. These use mechanical comparators. The translation of the linear from of the work piece is accomplished by referencing the caliper jaws to a measurement scale by mechanical mean. The work piece variation is assessed by a gear-train and lever (mechanical) translation technique that utilizes principles of amplification. Calipers utilizing electronic technology to assess and display dimensions are also available.

Many modern calipers are fitted with computerized storage and display unit; which can chart and interpret assessed variation over time.

Surface Plates

The foundation of all geometric accuracy and indeed of all dimension measurement is the flat plane, which is usually provided in the workshop and inspection laboratories by the surface plate. Most surface plates are rectangular, having 4:3 length: width ratio. Square or round plates are geometrically most correct for creating a flat plane by the three plate method.

The commonly used cast iron surface plates have special ribbing and structural formations under the surface so that these will be as light in weight as possible and yet will not warp or settle after the castings have aged. In fact, the true plane of the surface is not machined and "scarped" until it has really aged and all internal stresses have subsided. During the last few years true, flat, smooth surface have been polished on granite blocks and granite surface plates have found easy popularity. The iron plate has the advantage of allowing a certain amount of wringing and surface or edges are not readily chipped out if something is dropped on it. The glass plate will shatter, or break and be destroyed if heavy pieces are dropped on it.

Surface plate forms the basis of measurement and is a must for metrology laboratory where inspection work is carried out. It provides a datum or reference surface for measurements. It is also used to check the flatness of another surface. It has got a flat surface and all other measuring instruments having flat base can be placed on the surface plate. The big surface plate generally rests on four levelling screws with the help of which its surface can always be kept horizontal. This is invariably made of cast iron, and in manufacture, is artificially aged to set free from all internal stresses. After this, its dimensions will never change with time. At the bottom it is strengthened by ribs. The top surface is scraped to true flatness. A wooden cover is always placed over it, when not in use. When in use the whole surface of the plate must be used and not the centre or corners in order to have uniform wear. Also the grease must be applied to top surface when not in use for long time. A regular check for flatness must be made after regular intervals.

It is well known fact that all precision measurements start from flat surfaces and that without the aid of an accurate plane surface it is impossible to do any precise mechanical work and that the surface plate, whose upper face is flat within close limits of accuracy, is of fundamental importance for precision measurements. The surface plate serves two important purposes. *Viz.* it is used as datum plane for taking accurate linear and angular measurements, and with the aid of suitable marking compound, it serves for testing the degree of flatness of other surfaces. For very accurate work it is very essential that the form and dimensions of a surface plate must be such that distortion or deflection is reduced to a minimum even under comparatively heavy loads. For this reason it is customary to reinforce the plate proper by the provision of a frame and ribbing of considerable depth. The standard available sizes of the surface plats vary form 100 * 100 mm to 2000 * 1000 mm in about 13 ranges. Generally surface plates up to size of 1000 * 630 are provided with three supporting feet and above it, either 3 or 4 feet. If surface has more than 3 feet, the supporting surface on which the feet rest; a special stand or bench should be provided with adjustment to ensure that all the feet bear evenly. Moreover, it is also desirable that the bearing surface should be smoothly machined in order to permit slight movement with change in temperature. It is recommended to provide a projection of at least 25 mm on all sides of the surface plates measuring 250 * 250 or over. The undersides of these projections should be machined properly and reasonably flat to facilitate the clamping of work or measuring instruments. The threaded holds can also be provided in the plate without affecting the accuracy if the drilling and tapping operations are carried out before the surface in finished.

The four edges of precision plates are usually finished straight and square to each other within the accuracy of 0.08 mm per metre. The upper surfaces of plates are finished by scarping operation. The ideal of the accuracy of the flatness of the top surface can be had from the fact that the maximum permissible error from mean true plane (for a surface plate of diagonal 150 mm) is allowed between ± 0.002 to ± 0.005 mm; and for surface plate of diagonal 2000 mm, maximum permissible error from mean true plane is allowed to vary between ± 0.02 to ± 0.04 mm. Further in additional to the conforming of above tolerances of the top surface of the plate, bearing area of any local area from the mean true plane of that local area should not deviate by more than 0.01 mm per metre to 0.025 mm per metre. It is also very essential that the proportion of the area of the surface plate represented by high spots, on which another flat surface would rest, should be comparatively small and should not be more than 10 to 20%.

Material for Surface Plate

The surface plates should be made from a material which will provide a high degree of rigidity and freedom from warping and capable of taking a high finish and the surface must be resistant to wear. Most commonly used material for this purpose is either the plain or alloyed closed-grained cast iron of good quality. After casting and rough machining and prior to finishing operation, the plates should be allowed to age either naturally by allowing a suitable period of time to elapse or by proper heat treatment in order to relieve the internal stresses.

The composition of most commonly used plain cast iron and alloy cast iron are:

Plain Cast Iron		***Alloy Cast Iron***	
Total carbon	...3 to 3.5%	Total carbon	...2.8 to 3.2%
Combined carbon	...0.4 to 0.7%	Silicon	...0.8 to 1.5%
Manganese	...0.5 to 1.2%	Manganese	...0.6 to 1.0%
Silicon	...1.0 to 1.2%	Sulphur	...0.12% (max)
Sulphur	...0.15% (max)	Phosphorus	...0.33% (max)
Phosphorus	...1.2% (max)	Nickel	...1.4 to 1.6%
		Chromium	0.4 to 0.6%

The heat treatment of the plates made of above materials is carried out as follows:

The plates are placed in an annealing furnace and heated upto 450 to 500°C and maintained at this temperature for about 3 hours or longer depending on the mass of the casting. The castings are protected from the direct heat of the flames by means of suitable baffle plates and heating should be as uniform as possible throughout. In case of small castings, more uniform heating is obtained by packing the casting in iron fillings in boxes. Then plate is allowed to cool gradually in the furnace itself without opening the doors. It is preferable to hold the temperature at 400°C for about an hour during cooling.

Apart from cast iron, surface plates are also made of non-metallic substances, *viz.* granite and glass. Metallic objects like height gauges, squares etc. can be easily slid over their surface. Their advantages are: These are more rigid for same depth. Corrosion is virtually absent. A big problem with C.I. surface plate is that any damage to surface throw up a projecting burr but in this case it causes indentation only.

Manufacture of Cast iron Surface Plates

It is not uncommon for a cast iron part to lose its straightness, etc. over a period of time if certain precautions are not taken during its manufacture and use. The composition of the iron is the first thing to be taken care of since the proper composition improvement mach inability, wear qualities and the quality of the scraped surface. It is also important to remember that stability of the iron depends mostly on slow, uniform cooling in the mould after casting, which in true, depend on the moulds of proper design, and on allowing sufficient time for complete cooling before the casting is exposed to air.

Angle Plate

This is an accessory needed with surface plates for measurement purposes. The two surfaces of it are perpendicular to each other. It is available in various designs. Box squares are also sometimes used in place of angle plates. It is made of cast iron and its surfaces are scraped.

Cast iron angle plates are widely used for workshop and inspection purposes. These are generally made from close grained cast iron and have a minimum hardness of 180 HB. After being cast and rough machined, the plates are given suitable heat treatment to relieve internal stresses before being finished. The heat treatment operation is carried out by placing the angle plates in an annealing furnace and heating up to a temperature of about 550°C and maintaining at this temperature for about 3 hour or more depending upon the mass of the casting.

The castings are protected from the direct heat of the flames by suitable baffle plates and in small plates more uniform heating is achieved by packing the casting in iron fillings in boxes. The castings are then cooled at a rate of 300°C per hour within the furnace itself without opening the door. The material of the angle plates should be sound and free from blow holes and porous patches except for minor defects which can be repaired by plugging the material of the similar composition. General no sharp edges are allowed in the plates.

The angle plates are available in two grades depending upon their accuracy, *i.e.* grades 1 and 2. the angle plates should be flat within the tolerances specified for each grade (given on next page) and the departure from flatness can be tolerated only in the nature of concavity and convexity. It is the common practice to mark each angle plate legibly and permanently with the manufacturer's trade marks, the grade and other identification marks in such a way that the accuracy of the plates is not affected. Usually all the finished surfaces and edges of the plates are protected against climatic condition by applying suitable corrosive preventive preparation on them.

Size no.	L	B	H	T(minimum) For plates having no T-slots		R(Min) (for Grade 2 only)
				Webbed	Unwebbed	
1	125	75	100	13	16	8
2	175	100	125	16	19	8
3	250	150	175	22	25	10
4	350	200	250	29	35	10
5	450	300	350	32	38	13
6	600	400	450	35	-	13
7	700	420	700	35	-	15
8	1000	600	1000	40	-	20
9	1500	900	1500	50	-	25
10	2000	900	2200	50	-	25

Standard Sizes of Angle Plates

The general view of the angle plate and its important dimensions for various size numbers as per IS: 2554¯1963.

Grade 1 angle plates of size 3 to 6 when not finished on their interior faces are webbed. Grade 2 angle plates may or may not be webbed. When webs are provided, they are set in slight from the ends of the plates.

It is recommended that grade 1 angel plates on all their exterior faces, edges and interior faces be finished by either grinding operation or by hand scraping and conform to the accuracies specified in tabular from below. In this case the area of high spot (bearing area) should not be less than 20%. For grade 2, all the exterior surfaces are finished by planning or milling operation.

V-Block

These are widely used for workshop and inspection purposes for checking out the roundness of cylindrical work pieces and for marking centres accurately, etc. Generally the angle of V is 90° and these are available in wide variety of shapes. Sometimes clamps are also provided which bridge the vee to secure the work. These are also supplied as matched pairs in which case two V-blocks are of similar size and of the same grade of accuracy. In the case of V-blocks the working surfaces are flanks of vees, base and faces, top and side face.

Depending upon the accuracy, IS: 2949¯1964 specifies the v-block into two grades, *viz.* grade A and grade B. As in the case of surface plates, here also the tolerance on flatness is defined as the maximum permissible distance separating two imaginary parallel planes, within which the surface under consideration can just be closed. Similarly the tolerance on squareness is defined as the maximum permissible distance separating two imaginary parallel planes, (which are perpendicular to the datum face of the part in question) within the surface under consideration can just be enclosed.

V-blocks are furnished by hand scraping process till they have a bearing area of not less than 20%. Any departure from flatness in case of vees should be a convexity and on other surfaces it should be a concavity. In a direction parallel to the axis of the vee the departure from flatness should be a concavity and not convexity.

Various Forms of Vee Blocks

Two types of V-blocks commonly used. A V-block having two vees, one of them being deeper and wider than the other. A larger vee block having one vee only. For special purposes such as checking triangle effects or taps and other three-fluted tools, 120 degree included angle vee-blocks are also available. The vee block with the clamp.

The major purpose of the vee-blocks is to hold cylindrical pieces, or move to the point, to establish precisely the centre line or axis of a cylindrical piece. In using a vee-block, it is very essential that the cylindrical piece should rest firmly on the sides of the vee and not on the edges of the vee. Before using a vee-block, first check it visually, or with the finger nail, for dent, edge nicks, scratches and burrs which might either damage the surface plate or cause the vee-block to tilt imperceptibly from true flat or vertical.

Any Vee-block should be checked periodically for basic accuracy. If it has rusted, or worn, or warped a little; or if it were made inaccurately; its four sides and two ends may not be truly parallel or square to each other. The vee channel may be out of parallel. Constant use with the same size cylindrical work pieces wears hollows in the vee-sides. Some of possible types of inaccuracies discoverable in a vee-block are indicated by the dotted lines with a dial indicator.

Cast Iron Vee Block

Brief Specifications. The Vee Blocks are made of close grained high quality cast iron. They are accurately ground and numbered in pairs, so that the Vee-Grooves of the same numbers are always in alignment.

a) Accurately ground to 90 degrees square.

b) Straightness within ± 0.001 mm per 20 mm length.

c) Symmetisation of Vs within 0.002 mm per 20 mm length.

Cast Iron Surface Plates with Handles

Brief Specifications. Cast Iron Surface plates with handles are made of close grained high quality iron.

Casting strains are eliminated either by natural ageing or by artificial seasoning.

Grade A: Handscraped and spotted.

Accuracy: ± 0.002 mm per 240 mm length.

Grade B: Ground handscraped.

Accuracy: ± 0.008 mm per 300 mm length.

Bench Centres

Bench centre consist of a rigid cast iron base made from close grained cast iron having minimum hardness of 180 HB (free from distortion, porosity and other casting defects). The base is provided with suitable T-slots for the attachment of the centre holders. The centre holders can be located in any desired position. The male or female centres are adjustable and can be locked in any position along the ground vees of the centre holders. These are used for checking eccentricity of rotary components between centres.

These are available with heights of centres 125, 160, 200, 250 and 300 mm. for easy loading and unloading, one of the centre is spring loaded. The distance between the centres may vary from 500 mm to 1500 mm depending on the height of the centres. The coaxially of centres and parallelism of axis of centres with respect to guide ways should be ensured within close tolerances.

The permissible deviation in parallelism of the axis of centres with respect to guide ways, as per IS 1980—1978 are 0.01 mm per 300 mm, and 0.015 mm/300 mm respectively leaning towards the free end of mandrel for centre height of 125—160 mm and 200—300 mm respectively. The corresponding permissible errors in coaxially of centres are 0.01 mm and 0.015 mm respectively over any length of 300 mm.

Straight Edges

These are used for checking the straightness and flatness of parts in conjunction the surface plates and spirit levels. These may be made of steel or cast iron. Steel straight edges are available up to 2 m length and may be rectangular in section with beveled edge. C.I. straight edges are made up to 3 m length and widely used for testing machine tool side ways. They are heavily ribbed and bow-shaped (camel back construction) to prevent distortion. These are provided with feet for rest when they are idle to prevent distortion. Feet are placed at points of minimum deflection.

The straight edges are classified as follows:

i- Tool-maker's straight edge.

ii- Wide-edge straight edge.

iii- Angle straight edge.

iv- Box straight edge.

Tool maker's straight edges have the highest accuracy and are available with various cross-sections, from one to four working edgesand length from 75 mm to 175 mm. Straight edges with a single edge are used for checking straightness by sight test. The tests consist of applying the straight edges along the full length of the surface (whose straightness has to be checked) against a bright back-ground. The absence of light between the straight edges and the surface indicates the straightness of element. Deviation of 1 or 2 microns from the straightness can be detected by the eye with some practice.

Flatness of a surface may be tested with single edged straight edge by applying it in different direction at different places on the tested surface. The flatness of the surface is judged by the light showing through. A surface can also be tested by means of three and four edges type straight edges by applying a light coat of Prussian blue on the working edges and then drawing the straight edge across the tested surface. Trace of marking compound are rubbed in this way on the tested surface and due to irregularities in the surface it will be coated in spot with different density. All the high spots are painted more densely and the low spots are partly painted. The surface is then scraped or ground until a uniform distribution of spots on the whole surface is obtained after subsequent tests.

Straight edges with a wide working edge are used for testing large surface or surfaces with large intermediate gaps or recesses. Such straight edges may have a length up to 0.6 metre, and have a form of rigid beam or frame in order to possess ample of rigidity to maintain the straightness.

Angle straight edges are used for checking surfaces positioned at an angle in relation to each other. Such straight edges have a triangular or trapezoidal cross-section and are from 250 to 1000 mm in length and usually have handles at their ends to facilitate their application. Surfaces are usually tested with a marking compound as described above. The recommended lengths of straight edges for bow shaped and I-section type are as follows.

Lengths of bow-shaped straight edges (L): 300, 500, 1000, 2000, 4000, 6000 and 8000 mm.

Lengths of I-section straight edge (L): 300, 500, 1000, 2000, 3000, 4000 and 5000 mm.

Bow-shaped straight edge has two feet of the same width as the working surface and these are so located as to allow the straight edge to be used in 3 positions *viz.* resting upon the feet; with the working surface downward; and lying on either side. The accuracy requirements specified for straight edge are applicable for all these three positions.

Usually two grades of accuracy (Grade A and Grade B) are specified. The working surface and supporting feet of grade A straight edges are finished by scraping, and the working surfaces of grade B and the side faces of all straight edge, are finished by smooth machining.

The proportion of bearing area of the working surface should not be less than 20% and 10% respectively for grades A and B ; High spots should be uniformly distributed and the percentage of bearing area should not be so high as to cause wringing. All sharp edges should be removed and the unmachined parts painted.

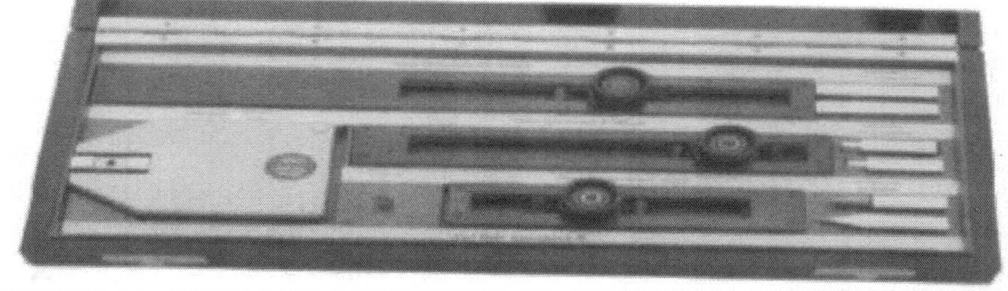

Spirit Levels

Spirit levels are used for measuring small angle or inclinations and also enable the position of a surface to be determined with respect to the horizontal.

A spirit level consists of a sealed glass tube, ground on its inside surface to a convex form with a large radius of curvature R. A scale is engraved on the top of the tube.

The tube is nearly filled with ether such that only a small volume remains at the top part of the tube which contains ether vapours in the form of a bubble. The machinist's level consists of a body with a flat base surface and the glass tube (described above) mounted in the upper part of the body.

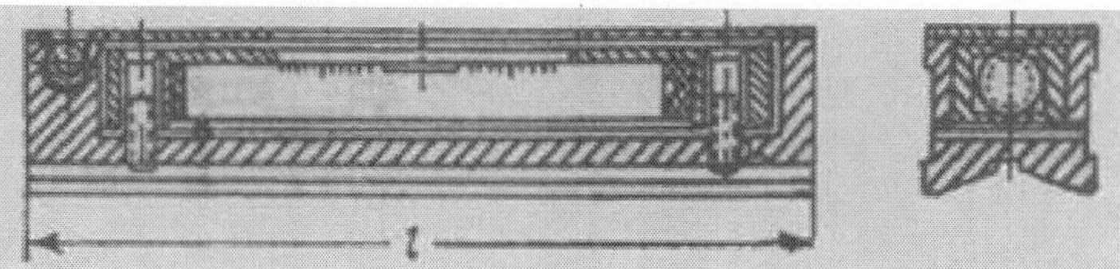

For checking the vertical surfaces, a frame level is employed.

The side edges of the frame level are made strictly square with the base. A glass tube filled with ether is mounted in the base.

For checking the vertical surface, the side edge of the frame is placed into exact contact with the surface and reading of the bubble noted down.

The position of the frame level in transverse direction is checked by another less accurate tube.

Principle

The glass tube is set in the base and adjusted in such a way that when the base is horizontal the bubble rests at the centre of the scale which is engraved on the glass. When the base of the level is moved out of the horizontal, the bubble tries to remain at the highest point of the tube and thus moves along the scale.

The relation between the movement of bubble and other conditions involved are as follows:

B is the top of the tube radius and the position of the bubble when the base is at *OA* (horizontal). If the base is tilted through an angle α and base occupies position *OA*', the bubble will move a distance 1 to *B*', where angle *BOB*'= α. If R is the radius of the tube then arc $l=R\ \alpha$.

If L is the length of the base and h is the difference in height between its ends, then for small value of $h, h=L\alpha l=Rh/L$ and if α is taken in seconds, then $l=R\alpha/206{,}265$ where (One radian equals 206,265 seconds of an arc). From above relation it is obvious that the sensitivity of the level increases as R increases.

The scale spacing or the distance between adjacent graduation is generally about 2 mm and thus for $R=206m$.

$$\alpha=2206{,}265/206{,}000\text{H} \approx 2 \text{ seconds.}$$

The inclination of 2" causes bubble movement of 2 mm. This is the sensitive spirit level and is recommended for research laboratory.

For highly precise shop measurement, spirit levels with scale division value of 4 to 10 are employed. For ordinary purposes, scale division values of the order of 10 to 40 are sufficient.

It may be noted that spirit levels are very sensitive to variation in the temperature of their surroundings, since they change the tension of ether vapours in the tube. As such the spirit level should be used in controlled temperature condition, or else precise spirit levels equipped with various devices that compensate for errors due to temperature change be used.

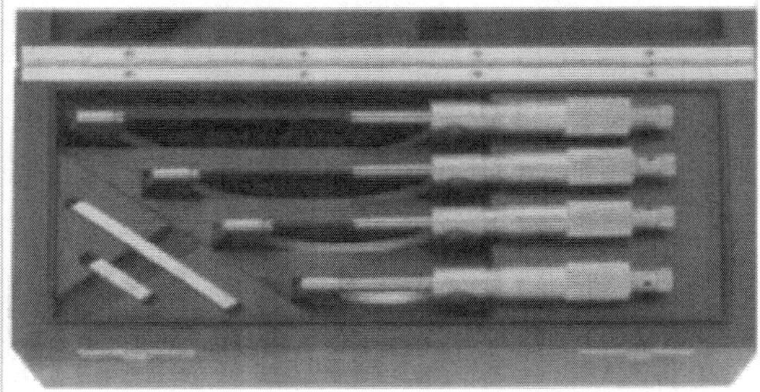

Types of Spirit Levels

According to *BS* 1958, three types of spirit levels are recommended. *Type 1.* This type of level has an unrelieved flat base of steel, hardened and lapped. The advantage of lapped is that the effective length of the level can be varied by wringing two guage blocks on the base at the desired distance apart. The bas length of spirit level varies between 100 to 200 mm.

Type 2. Level is mounted in cast iron or steel body, (with base length 250-500 mm) having abase formed with feet bearing surfaces at the two ends (middle portion being relieved). The bearing surface may be plain or contain a longitudinal 120 degree vee groove for use on cylindrical surface, in which case a short cross-level or small circular level is provided.

Type 3 level is a square block level about 200 mm square and made of cast iron. The four bearing surface are flat and may have the middle portion relieved. Alternatively, the base and one adjacent-surface may contain a longitudinal 120° vee groove for use on cylindrical surface in which case a short cross-level is provided.

Electrical Level

This is a type of precision level in which the usual spirit vial is replaced by a pendulum whose displacements are electrically amplified and shown on a separate metre.

Combination Set

This is the most adaptable and commonly used non-precision instrument to be used in layout and inspection work.

The combination set consist of scale, squaring-head, protractor and centre-head. It consists of a heavy scale, which is grooved all along its length. It is on this groove that sliding squaring head is fitted. One surface of the squaring head is always perpendicular to the scale and it can be adjusted at any place by a locking blot and nut. The squaring head also contains a spirit level which is used to test the surface for parallelism. For laying out dovetails an included angle is also mounted on the scale. It can also slide to any position and be locked there. A scribing point is also inserted into the

Radius Gauges

Wherever two surfaces meet, a radius is generally given to avoid stress concentration. The radius also plays an important role in case of heat treatment of parts. Thus the proper radius must be maintained and for checking it, radius gauges are employed. If the curved surface is external, i.e. male radius it is simply called a 'radius'. If curved surface is internal, i.e, female radius, it is called a 'fillet'. Radius gauges consist of sets of blades on each of which is stamped the corresponding radius. On one side we have external radius and on the other side internal radius, so that it may be suitable for fillets as well a radius. Radius gauges are used with proper light, placing the work to be gauged between light source and eye. It is only with the passage of light between gauge and work that radius can be gauged properly.

Fillet or Radius Gauges

It contains external and internal radius on each leaf permitting both concave and convex surfaces to be measured. The leaves are specially shaped so that they can be used in any position at any angle to measure fillets and radius in corners or against shoulders.

Figure: *Fillet or radius gauges*

Feeler Gauges

These are used to measure the width of the gap between two parallel flat faces e.g., in gauging of the clearance between the piston and cylinder. A feeler gauge consists of a narrow strip of sheet steel made to a given thickness. The complete set consists of a number of gauging blades of different thickness assembled together. Their working depends entirely on the sense of feel. The feeler blade should neither be forced between the surfaces nor should it slide freely, rather the correct blade will give a characteristic 'gauge fit' type of feel. If necessary, two blades may be joined together for noting any dimension.

Feeler gauges generally comprise of a series of gauging blades of different grades and thicknesses, from 0.03 to 1 mm assembled in protective sheath. The blades are made of heat treated bright polished tool steel having a tensile strength of about 170 kgf/mm2 for thickness up to 0.5 mm and about 70 kgf/mm2 for thickness 0.5 mm and above. Generally the blades are available in overall length of 100 mm. these are 12 mm wide at heel and are tapered for the outer part of their length so that width at the tip is approximately 6 mm. these are hinged in the sheath on a screw and nut of such a design that the blades are removable. The nut is in the form of a bush, passing through both sides of the sheet and forms a hinge upon which the blades may be rotated. The sheath is so designed as to fully protect the blades when not in use.

Each blade is legibly and permanently marked with the nominal thickness and grade. These are available in two grades depending upon the accuracy. The various blades should be assembled in such a way that each thin blade is given the maximum protection by being inter- leaved between two thicker blades. The maximum variation in the thickness of a blade should not exceed 0.04 mm for blades upon and including 0.3 mm thick and 0.006 mm for blades over 0.3 mm thickness.

The recommended set by Bureau of Indian Standards which has been so devised as to furnish sets of the greatest utility with a minimum number of blades is given below:

Set No.	*No. of blades in the set*	*Thickness of blades in mm*
1	8	0.03 to 0.1 in steps of 0.01
2	9	0.03, 0.03, 0.04, 0.04, 0.05, 0.05, 0.06, 0.07, 0.09
3	16	0.03, 0.04, 0.05, 0.06, 0.07, 0.08, 0.09, 0.1, 0.15, 0.2, 0.25, 0.3, 0.35, 0.4, 0.45, 0.5
4	11	0.03, 0.04, 0.05, 0.06, 0.07, 0.1, 0.15, 0.2, 0.3, 0.4, 0.5
5	14	0.05, 0.06, 0.07, 0.08, 0.09, 0.1, 0.15, 0.2, 0.25, 0.3, 0.4, 0.5, 0.75, 0.1
6	11	0.05, 0.06, 0.07, 0.08, 0.09, 0.10, 0.15, 0.2, 0.4, 0.75, 1.0
7	11	0.5, 0.55, 0.6, 0.7, 0.75, 0.8, 0.85, 0.9, 0.95, 1.0

Angle Gauges

These are similar to feeler gauges. A set of angle gauge consists of 18 blades with their ends cut at various angles from 2 to 45 degrees.

Engineer's Taper, Wire and Thickness Gauge

This is a very handy device which contains leaves for taper measurements, wire diameter measurements and thickness of small gaps.

Pitch Screw Gauge

It is used to quickly determine the pitch of various threads by matching the teeth on the leaves with the teeth on the work.

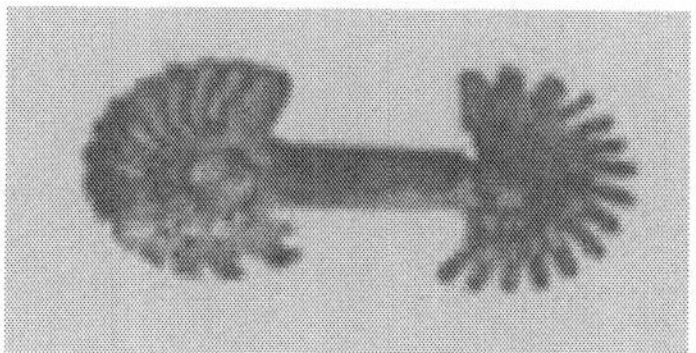

Figure: *Pitch screw gauge*

Precision Linear Measurements

As modern production is concerned with Interchangeable products, a great precision dimensional control is required in industry. For this to be achieved, importance of precision measuring instruments can be well visualized.

Characteristics of Precision Measuring Instruments

In order that precision measuring instruments function properly, they should possess the following characteristics: (a) high degree of sensitivity (b) high degree of accuracy and (c) minimum inertia in the moving parts of mechanism. Sensitivity of the instrument may be defined as the rate of displacement of the indicating device with respect to the quantity to be measured. Literally, sensitivity is defined with reference to the least change in the measured quantity which will cause an observable change in the instrument reading. In instrument design an attempt must be made that sensitivity remains constant throughout the range of measured quantity.

The measure of accuracy of an instrument can be had by the amount of correction to be made to the instrument reading. The accuracy will be high, if calibration is proper and less wearing of parts takes place. The effect of inertia of moving parts is to make the instrument sluggish. All those instruments which depend on linkage and mechanical system, displacement of fluid, diaphragm etc. are subjected to disadvantages of inertia. Thus imperfect elasticity of the diaphragms and spring are causes of inertia in instruments. However instruments based on the principle of optics are entirely free from inertia defects.

In addition to these three characteristics, the measuring instruments must have freedom from variance. Variance is defined as the range of variance in instrument readings obtained from repeated measurements of same quantity. It is inherent in the instrument and depends upon the quality of instrument which in turn depends upon the quality of machines from which it is manufactured and quality of workmanship.

Vernier Instruments

The principle of vernier is that when two scales or divisions slightly different in size are used, the difference between them can be untilised to enhance the accuracy of measurement. The vernier caliper essentially consists of two steel rules and these can slide along each other. One of the scales, i.e., main scale is engraved on a solid L- shaped frame. On this scale cm graduations are divided into 20 parts so that one small division equals 0.05 cm. one end of the frame contains a fixed jaw which is shaped into a contact tip at its extremity.

The three elements of vernier caliper, viz. beam, fixed jaw, and sliding jaw permit substantial improvements in the commonly used measuring techniques over direct measurement with line graduated rules. The alignment of the distance boundaries with the corresponding graduations of the rule is ensured by means of the positive contact member (the jaws of the caliper gauges). The datum of the measurement can be made to coincide precisely with one of the boundaries of the distance to be measured. The movable jaw achieves positive contact with the object boundary at the opposite end of the distance to be measured. The closely observable correspondence of the reference marks on the slide with a particular scale value, significantly reduces the extent of read- out alignment errors.

A sliding jaw which moves along the guiding surface provided by the main scale is coupied to a vernier scale. The sliding jaw at its left extremity contains another measuring tip. When two measuring tip surfaces are in contact with each other, scale shows zero reading. The finer adjustment of the movable jaw can be done by the adjusting screw. First the whole movable jaw assembly is adjusted so that the two measuring tips just touch the part to be measured. Then lock nut B is tightened. Final adjustment depending upon the sense of correct feel is made by the adjusting screw. The movement of adjusting screw makes the part containing locking nut A and sliding jaw to move, as the adjusting screw rotates on a screw which is in a way fixed to the movable jaw. After final adjustment has been made, the locking nut A is also tightened and the reading is noted down. The measuring tips are so designed as to measure inside as well as outside dimensions.

Reading the Vernier Scale

For understanding the working of vernier scale let us assume that each small division of the main scale is 0.025 unit. Say, the vernier scale contains 25 divisions and these coincide exactly with 24 divisions of main scale. So now one vernier division is equal to 1/25 of 24 scale divisions, i.e., $1/25 \times 24 \times 0.025 = 0.024$ unit. Therefore, difference between one main scale small division and one vernier division (least count of the instrument) equals 0.025- 0.024, i.e. 0.001unit. It means if the zero of main scale and zero of vernier concide, then the first vernier division will read 0.001 unit less than the 1 small scale division. Second vernier division will read 0.002 unit less than 2 small scale divisions and so on. Thus if zero vernier scale lies in between two small divisions on main scale its exact value can be judged by seeing as to which vernier division is coinciding with main scale division.

Thus to read a measurement from a vernier caliper, note the units, thenths and fortieths which the zero on the vernier has moved from the zero on the main scale. Note down the vernier division which concides with a scale division and add to previous reading the number of thousands of a unit indicated by the vernier divisions. E.g, reading in the scale is 3 units + 0.1 unit + 0.075 unit + 0.008 unit = 3.183 units. When using the vernier caliper for internal measurement the *width of the measuring jaws must be taken into account.* (Generally the width of easuring jaw is 10 min for Metric System).

Types of Vernier Calipers

According to IS: 3651 ..... 1974 (Specification for vernier caliper), three types of vernier calipers have been specified to meet the various needs of external and internal measurements upon 2000 mm with vernier accuracy of 0.02, 0.05 and 0.1 mm. the three types are called types A, B, C and have been shown respectively.

All the three types are made with only one scale on the front of the beam for direct reading. Types A has jaws on both sides for external and internal measurements and also has a blade for depth measurements. Type B is provided with jaws on one side for external and internal measurements. Type C has jaws on both sides for making the measurements and for making operations.

All parts of the vernier calipers are made of good quality steel and the measuring faces hardened to 650 H.V. minimum. The recommended measuring ranges (normal sizes) of vernier calipers as per IS 3651 ..... 1974 are 0 ...... 125, 0 ..... 200, 0 ..... 250, 0 .....300, 0 ..... 500, 0 ..... 750, 0..... 1000, 750 ..... 1500 and 750 ...... 2000 mm.

Graduation on beam at every 1 mm and each 5 mm line extended and each 1 cm line is numbered. On vernier scale there are 20 divisions within a distance of 19 mm and 19 mm= 19 divisions of main scale.

On type A, scale serves for both external and internal measurements, whereas in case of types B and C, the main scale serves for external measurements and for marking purposes also in type C, but on types B and C internal measurements are made by adding width of the internal measuring jaws to the reading on the scale. For this reason, the combined width for internal jaws is marked on the jaws in case of types B and C calipers. The combined width should be uniform throughout its length to within 0.01 mm.

Graduations on beam are at every 1/2 m and every alternate mm lines are extended and numbered 2, 4, 6, 8.

On vernier scale, there are 10 divisions within a distance of 9.5 mm and 9.5 mm = 19 divisions of main scale.

The beam for all the types is made flat throughout its length to within the tolerances of 0.05 mm for nominal lengths upon 300 mm, 0.08 mm from 900 to 1000 mm, and 0.15 mm for 1500 and 2000 mm sizes, and guiding surfaces of the beam are made straight to within0.01 mm for measuring range of 200 mm and 0.01 mm every 200 mm measuring range of larger size. The measuring surfaces are given a fine ground finish. The portion of the jaws between the beam and the measuring faces are relived.

The fixed jaw is made an integral part of the beam and the sliding jaw is made a good sliding fit along with the beam and made to have seizyre- free movement along the bar. A suitable looking arrangement is provided on the sliding jaw in order to effectively clamp it on the beam. When the sliding jaw is clamped to the beam at any position within the measuring range, the external measuring faces should remain square to the guiding surface of the beam to within 0.003 mm per 100 mm. the measuring surfaces of the fixed and sliding jaws should be coplanar to within 0.05 mm when the sliding jaw is clamped to the beam in zero position. The external measuring faces are lapped flat to within 0.005 mm. the bearing faces of the sliding jaw should preferably be relieved in order to prevent damage to the scale on the beam. Each of the internal measuring surface should be parallel to the corresponding external measuring surface to within 0.025 mm in case of type B and C calipers. The internal measuring surfaces are formed cylindrically with a radius not exceeding one- half of their combined width.

Errors in Measurements with Vernier Calipers

Errors are usually made in measurements with vernier calipers from manipulation of vernier caliper and its jaws on the workpiece. For instance, in measuring an outside diameter, one should be sure that the caliper bar and the plane of the caliper jaws are truly perpendicular to the workpiece's longitudinal centre line; i.e. one should be sure that the caliper is not canted, titled, or twisted. It happens because the relatively long, extending main bar of the average vernier calipers so readily tips in one direction or the other.

The accuracy of the measurement with vernier calipers to a great extent depends upon the condition of the jaws of the caliper. The accuracy and the natural wear, and warping of vernier caliper jaws should be tested frequently by closing them together tightly or sitting them to the 0.0 point of the main and vernier scales. In this position, the caliper is held against a light source. If there is wear, spring or wrap, a knock- kneed condition (a) will be observed. If measurement error on this account is expected to be greater than 0.005 mm the instrument should not be used and sent for repair.

When the sliding jaw frame has become worn or wrapped so that it does not slide squarely and snugly on the main caliper beam, then jaws would appear.

Where a vernier caliper is used mostly for measuring inside diameters, the jaws may become bowlegged or its outside edges worn down.

Care in Use of Vernier Calipers

These should not be treated or used as a wrench or hammer because these are not rugged instruments. They should be set down gently, preferably in the box and not dropped or tossed aside. They must be wiped free from grit, chips and oil. These should be brought to the work piece. The work piece should not be clamped in the caliper jaws and waved in air.

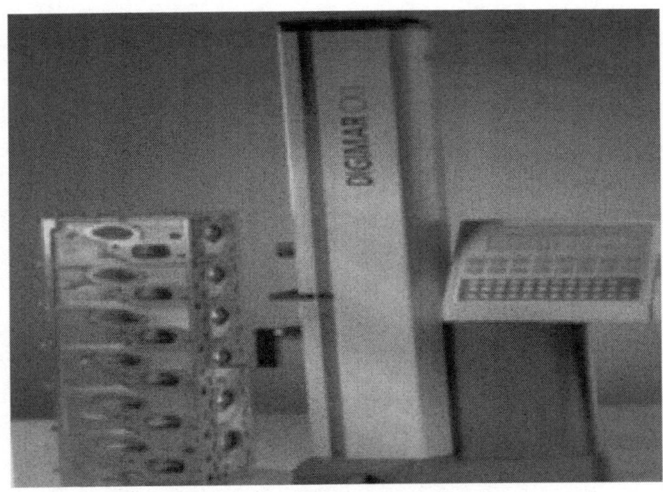

Precautions in the Use of Vernier Caliper

No play should be there between the sliding jaw on scale, otherwise the accuracy of the vernier caliper will be lost. If play exists then the gib at the back of jaw assembly must be bent so that gib holds the jaw against the frame and play is removed.

Usually the tips of measuring jaws are worn and that must be taken into account. Most of the errors usually result from manipulation of the vernier caliper and its jaws on the work piece.

In measuring an outside diameter it should be insured that the caliper bar and the plane of the caliper jaws are truly perpendicular to the work piece's longitudinal centre line. It should be ensured that the caliper is not canted, titled or twisted.

The stationary caliper jaw of the vernier caliper should be used as the reference point and measured point is obtained by advancing or withdrawing the sliding jaw.

In general, the vernier caliper should be gripped near or opposite the jaws; one hand for the stationary jaw and the other hand generally supporting the sliding jaw. The instrument should not be held by the over- hanging "tail" formed by the projecting main bar of the caliper.

The accuracy in measurement primarily depends on two senses, viz., sense of sight and sense of touch (feel). The short- comings of imperfect vision can however be overcome by the use of corrective eye-glass and magnifying glass.

But sense of touch is an important factor in measurements. Sense of touch varies from person to person and can be developed with practice and proper handing of tools. One very important thing to note here is that sense of touch is most prominent in the finger- tips, therefore, the measuring instrument must always be properly balanced in hand and held lightly in such a way that only fingers handle the moving and adjusting screws etc. if tool be held by force, then sense of feel is reduced.

Vernier caliper must always be held at short leg of main scale and jaws never pulled.

Vernier Height Gauge

This is also a sort of vernier caliper, equipped with a special base block and other attachments which make the instrument suitable for height measurements.

Along with the sliding jaw assembly, arrangement is provided to carry a removable clamp. The upper and lower surfaces of the measuring jaws are parallel to the base, so that it can be used for measurements over or under a surface.

The vernier height gauge is mainly used in the inspection of parts and layout work. With a scribing attachment in piace of measuring jaw, this can be used to scribe lines at certain distance above surface.

However dial indicators can also be attached in the clamp and many useful measurements made as it exactly gives the indication when the dial tip is just touching the surface. For all these measurements, use of surface plates as datum surface is very essential.

Specification of Vernier Height Gauges

For specifying the vernier height gauge, one has to specify clearly the range of measurement, the type of scales desired, and any particular requirements in regard to the type of vernier desired. Generally, all the parts of the height gauges are made of good quality steel, or stainless steel also in certain cases. At the time of fabrication, the blanks:

- Storage of upon 400 dimensions.
- Data retention on failure of electric power.
- Automatic entry of data as component is measured.
- Entry of data via the custom touch key- pad.
- Calculation of capability indicates for quick and easy assessment of machine/ plant or process capability.
- Easy identification of component features out of tolerance.
- Calculation of mean, standard deviation and range
- Assessment of 2- sigma, 3- signal, and 4- sigma confidence intervals
- Comprehensive print- out of statistical data including histograms.
- Each sequence of measurement is recorded in numerical order and can be recalled subsequently and printed out in the format required.

Micrometers

The micrometer screw gauge essentially consists of an accurate screw having about 10 or 20 threads per cm and revolves in a fixed nut. The end of the screw forms one measuring up and the other measuring tip is constituted by a stationary anvil in the base of the frame. The screw is threaded for certain length and is plain afterwards. The plain portion is called sleeve and its end is the measuring surface. The spindle is advanced or retraced by turning a thimble connected to the spindle.

The spindle is a slide fit over the barrel and barrel is the fixed part attached with the frame. The barrel is graduated in unit of 0.05 cm. i.e. 20 divisions per cm, which is the lead of the screw for one complete revolution. The thimble has got 25 divisions around its periphery on circular portion. Thus it sub- divides each revolution of the screw in 25 equal parts, i.e. each division corresponds to 0.002 cm.

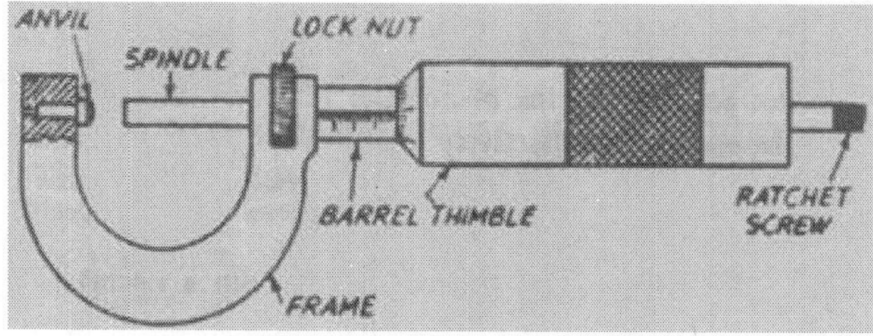

A lock nut is provided for locking a dimension by preventing motion of the spindle. Ratchet stop is provided at the end of the thimble cap to maintain sufficient and uniform measuring pressure so that standard conditions of measurement are attained. Ratchet stop consists of an overriding clutch held by a week spring. When the spindle is brought into contact with the work at the correct measuring pressure, the clutch starts slipping and no further movement of the spindle takes place by the rotation of ratchet. In the backward movement it is positive due to shape of ratchet.

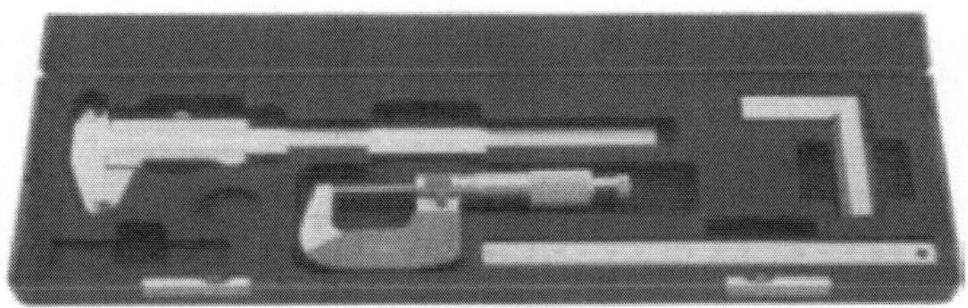

Adjusting for Wear of Threads, Wear of Measuring Surfaces, Spindle Locking Arrangement and Ratchet Stop Mechanism

In the external micrometer shown, an abutment or anvil is provided which occupies a fixed position in relation to the main nut, in which the spindle moves. The anvil and the main nut are carried by a bow-shaped frame. The main nut is fitted tightly into the sleeve or barrel which is integral with the frame. The thimble is permanently secured to the screw and is knurled on the outer surface.

The spindle having the measuring face at its end is tightly fitted into the screw. The spindle is thus rotated by rotating the thimble. It will be observed that the thimble extends over the end of the barrel so that threaded portions of the screw and the main nut are all the time completely enclosed. The nut and the thread portion of the screw are so proportioned as to ensure full length engagement in all positions of spindle.

The fixed index line is marked upon the barrel and the angular graduations around the left hand chamfered end of the thimble. It will be noted that in addition to the main nut three is a shorter nut also by its side with which the screw also engages. The adjacent faces of the main nut and the shorter nut are provided with small V- shaped teeth which from a clutch and prevent the shorter nut from rotating with the screw. In the recess between these two nuts is housed a light coil spring, the tendency of which is to force the two nuts apart and in this way any slight backlash between the threads in the nuts and the screw is automatically taken up.

Adjustment for wear threads. In addition to forming an abutment for the coil spring, the shorter nut also provides for adjustment in the event of wear on the threads. When the two nuts are originally assembled on the screw, the clutch teeth are meshed in a certain position and marks are provided on the outer surfaces to facilitate replacement in the event of the instrument being dismantled. The wear on the threads is compensated by withdrawing the screw and turning the smaller nut through one, two or more clutch tooth spaces from its original angular position and then bringing it in mesh with the main nut. In this way the threads on the smaller nut are slightly displaced. At the time of re- assembling, care must be taken to see that the clutch teeth on the two nuts are meshing properly and that the small nut is held in position until the screw has entered the main nut.

In another most commonly used method of compensating for wear on the thread, provision of an external tapered thread is made on one end of the main nut. The tapered thread is engaged by an adjusting nut having a parallel thread. When sufficient wear occurs between the two threaded portions, the adjusting screw is advanced so as to contact the bore of the main nut slightly and thus making it take up the clearance between the threads of the nut and the screw.

Ratchet Stop Mechanism

The object of the ratchet stop is to ensure that a certain maximum torque on the spindle is not exceeded and the sense of the feel of operator is eliminated and consistent readings are obtained. By providing this arrangement, when the spindle has engaged the work with a certain pressure, further rotation causes the ratchet merely to slip, no additional movement being imparted to the spindle. The ratchet stop mechanism is incorporated in the knurled extension provided at the end of the thimble. The knurled extension is free to rotate on its retaining stud. The inner face of this is provided with fine ratchet teeth, and through these teeth, and through these teeth and the spring loaded pawl the movement is transmitted to the spindle. As soon as the resistance to the motion of the latter reaches a certain value, the pawl is forced back against the pressure of the spring and ratchet slips.

Reading a Micrometer

In order to make it possible to read upon 0.0001 inch in micrometer screw gauge a vernier scale is generally made on the barrel. The vernier scale has 10 straight lines on the barrel and these coincide with exact 9 divisions on the thimble. Thus one small division in thimble is further sub- divided into 10 parts and for taking the reading one has to see which of the vernier scale division coincides with a division of the thimble. Accordingly the reading for the given arrangements will be

On main barrel: 0.120"

On thimble: 0.014"

On vernier scale: 0.0001"

Total reading = 0.1342".

Before taking readings, anvil and spindle must be brought together carefully and the initial reading noted down. Its calibration must be checked by using standard gauge blocks.

In metric micrometers, the pitch of the screw thread is 0.5 mm so that one revolution of the screw moves it axially by 0.5 mm. main scale on barrel has least divisions of 0.5 mm. the thimble has 50 divisions on its circumference.

$\therefore$ One division on the thimble = 0.5/ 50 mm= 0.01 mm.

If vernier scale is also incorporated then sub- divisions on thimble can be estimated upon an accuracy of 0.001 mm.

Reading of micrometer is 3.5 mm on barrel and 7 divisions on thimble

$= 3.5 + 7 \times 0.01$

$= 3.5 + 0.07 = 3.57$ mm.

Cleaning the Micrometer

Micrometer screw gauge should be wiped free from oil, dirt, dust and girl. When micrometer feels gummy and dust ridden and the thimble fails to turn freely, it should never be bodily dunked in kerosene or solvent because just soaking the assembled micrometer fails to float the dirt away. Further it must be remembered that the apparent stickiness of the micrometer may not be due to grit and gum but to a damaged thread or to a warped and sprung frame or spindle.

Precautions in Using Micrometer

In order to get good results out of the use of micrometer screw gauge, the inspection of parts must be made as follows: Micrometer should be cleaned of any dust and spindle should move freely.

The part whose dimension is to be measured must be held in left hand and the micrometer in right hand. The way for holding the micrometer is to place the small finger and adjoining finger in the U- shaped frame. The forefinger and thumb are placed near the thimble to rotate it and the middle finger supports the micrometer holding it firmly.

Then the micrometer dimension is set slightly larger than the size of the part and is slid over the contact surfaces of micrometer gently. After it, the thimble is turned till the measuring tip just touches the part and the final movement given by ratchet so that uniform measuring pressure is applied. In case of circular parts, the micrometer must be moved carefully over representative arc so as to note maximum dimension only. Then the micrometer reading is taken.

The micrometers are available in various sizes and ranges, and the corresponding micrometer should be chosen depending upon the dimension.

Errors in reading may occur due to lack of flatness of anvils, lack of parallelism of the anvils at part of the scale or throughout, inaccurate setting of zero reading, etc. Various testes to ensure these conditions should be carried out from time to time.

Inside Micrometer Calipers

This micrometer caliper has no U- shape frame and spindle. The measuring tips are constituted by the jaws with contact surfaces which are hardened and ground to a radius. One of the jaws is held stationary at the end second one moves by the movement of thimble.

A locknut is provided to check the movement of the movable jaw. This facilitates the inspection of small internal dimensions. This is not widely used as other types have various advantages over it. Its range is from 5 to 50 mm.

Method of Testing Internal Micrometers

The accuracy of the internal micrometer reading is checked in two parts, viz., checking the accuracy of the traverse of the measuring head and the checking of the accuracy of the overall lengths when the measuring head, set to zero, is associated with the various extension rods in turn.

The accuracy of the traverse of the measuring head is determined by clamping it in the Vee- block with its axis in line with a sensitive indicator and the feeler (with a flat contact face) of the indicator touching the rounded contact face of measuring head. The feeler is set initially such that it reads zero when the micrometer head is also reading zero and a slip gauge of convenient dimension has been introduced between the two faces.

The micrometer head is then set at same reading and correspondingly the slip gauge size is reduced. Any error in the reading is revealed by a corresponding departure of the indicator pointer from its initial position. In this way the accuracy is tested at various readings of the micrometer head.

In order to check both progressive and periodic errors, and in order that both the type of errors may be revealed, the reading in micrometer head of 0ÜÜÜ 25 mm range may be taken at following values 0, 2.4, 5.1, 7.7, 10.3, 12.5, 14.9, 17.6, 20.2, 22.8, 25.0 mm. the accuracy of the overall length is determined by using a vertical comparator. The indicator of the comparator is set at a height corresponding to that of the overall length to be measured, from the flat base. After this setting the internal micrometer is placed under it in the maximum position and the error noted down.

Micrometer Depth Gauge

It is used for measuring the depth of holes, slots and recessed areas. It has got one shoulder which acts as reference surface and is held firmly and perpendicular to the centre line of the hole. Here also for larger ranges of measurements, extension rods are used. The screw of micrometer depth gauge has range of 20 mm or 25mm. the length of the micrometer depth gauge varies from 0 to 225 mm. the rod is inserted through the top of the micrometer. The rod is marked after every 10 mm so that it could be clamped at any position. In using this instrument, first it must be ensured that the edge of the hole is free from burrs.

The scale here is calibrated in the reverse direction the accuracy again depends upon the sense of touch.

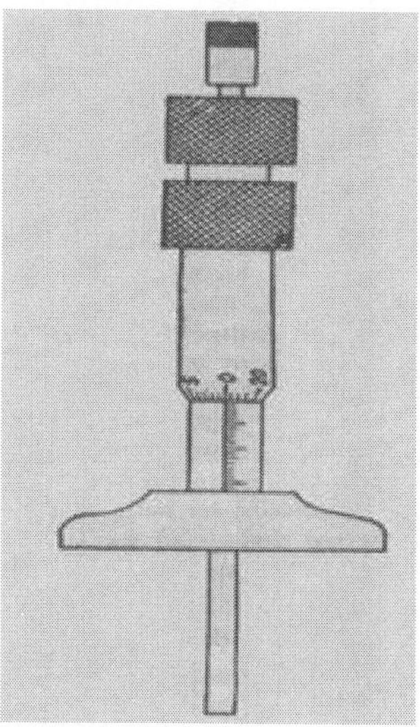

Thread Micrometer Caliper

This is just like an ordinary micrometer with the difference that it is equipped with a special anvil and spindle. The anvil has an internal vee which fits over the thread. The anvil in this case is not fixed but is free to rotate. Thus vee of the anvil can accommodate itself to any rake range of thread. The spindle on the other hand has a ground conical shape. When the conical spindle is brought into contact with the vee of anvil, micrometer reads zero. Different set of anvils are provided for different types of threads and the contact points of the micrometer are so designed that some allowance for thread clearance is always made.

Thread micrometer is used for the measurement of the pitch diameter but the accuracy is influenced by the helix angle of the thread.

*Screw thread micrometer caliper*its used for accurate measurement of pitch diameter of screw threads. The micrometer has a pointed spindle and a double V- anvil, both correctly shaped to contact the screw thread of the work being gauged. It directly reeds in terms of pitch diameter as the zero reading of the micrometer corresponds to the closed position of anvil and spindle when both are in perfect match with each other.

The angle of the V- anvil and the conical point at the end of the spindle corresponds to the included angle of the profile of the thread. The V- anvil is allowed to swivel in the micrometer frame so that it can accommodate itself to the helix angle of the thread.

The extreme point of cone is rounded so that it will not bear on the root diameter at the bottom of the thread, and similarly clearance is provided at the bottom of the groove in the V- anvil so that it will not bear on the thread crest. The spindle point of such a micrometer can be applied to the thread of any page provided the form or included angle is always the same. The V- anvil is however limited in its capacity; a number of different blocks being required to cover a full range.

*Outside Micrometer Caliper (*Box type with Sliding Anvil*)*

It is designed for fast, easy and precision measuring of large work having dimensions above 250 mm. generally six standards varying in length in step of 25 mm are furnished. It may be noted that it is impossible to check the zero readings by bringing the two measuring faces into contact. For this purpose a standard rod is employed. In order to increase the utility of instrument to cover wide range of measurement, standard rods of other dimensions are also provided which are fitted in place of fixed anvil. All these rods are provided with ebonite pads in order to prevent the transfer of heat from operator's hand.

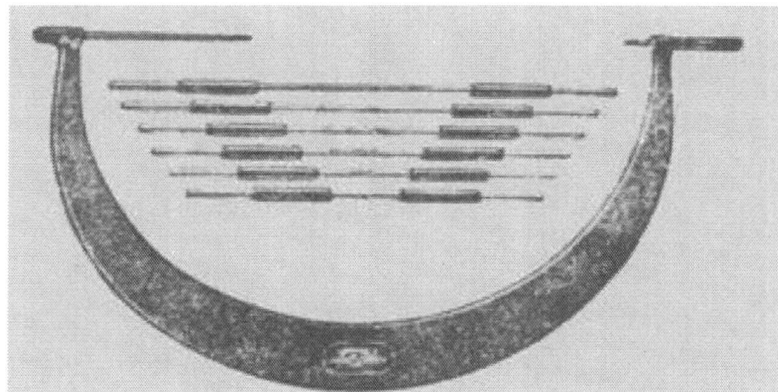

Figure: *Outside micrometer caliper.*

V- Anvil Micrometer Caliper

Special feature: Any out-of-roundness condition can be quickly checked in centreless grinding and machining operations.

Figure: *V- Anvil micrometer caliper*

Direct reading eliminates use of special fixtures. It can be used for measuring odd-fluted tapes, milling cutters, reamers etc.

In this micrometer, angle of Vee equals 60 degrees and the apex of the Vee coincides with axis of spindle. The zero reading of micrometer starts from a point where the two sides of Vee meet. It may now be seen that as the angle of Vee is 60°, the micrometer will measure a distance of $3d/2$ for a round piece of diameter d. It is thus simple matter to obtain the diameter from reading.

Blade Type Micrometer

It is ideally suited for fast and accurate measurement of circular formed tools, diameter and depth of all types of narrow grooves, slots, keyways, recesses, etc. it has non-rotating spindle which advances to contact the work without rotation.

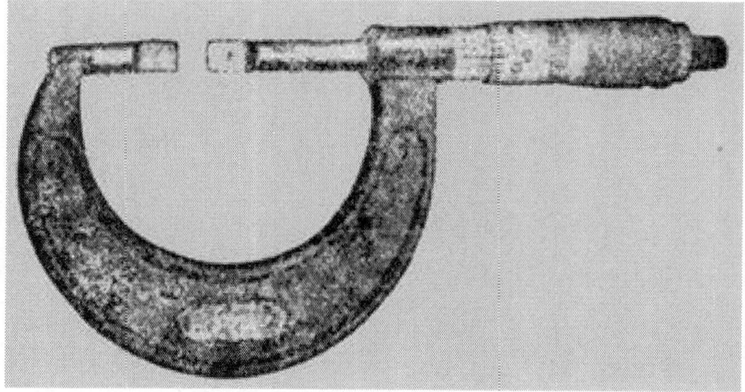

Figure: *Blade type micrometer.*

Micrometer for Measuring Thickness of Cylinder Walls

The ordinary micrometer can't be used for measuring the thickness of the wall of a tube, sleeve or bush because of the concavity of the internal surface. In the micrometers meant for this purpose, the fixed anvil is provided with a spherical measuring surface and the frame is cut away on the outside to permit of the anvil being introduced into tubes of diameters as small as 7.5 mm.

In another design shown, the anvil is made of cylindrical form, its axis being perpendicular to the axis of the spindle.

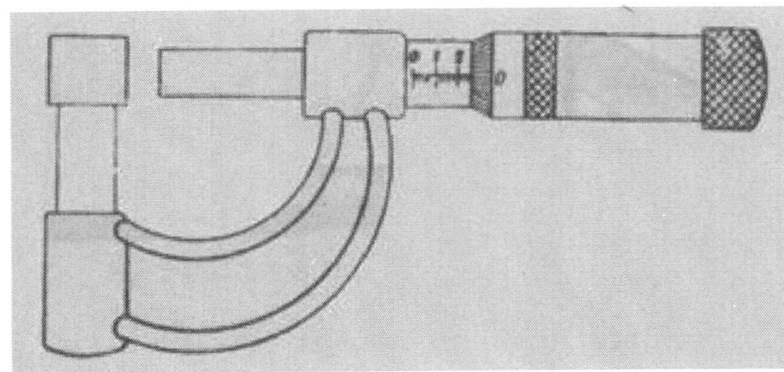

Dial Micrometer Caliper

The is equipped with a movable anvil whose slight axial movement is identical by a dial indicator, fixed into the frame. The arrangement is also provided to retract the anvil simply by pushing a button.

Due to provision of indicator, the correct readings can be obtained as the indicator will show reading the moment some measuring pressure is felt. It can be very useful for statistical quality control work as variations in size can be noted by the dial indicator.

Bench Micrometer

Special features:

- Super precision dial indicator for comparator measurements upon 1 micron (μm).
- High precision micrometer head with direct readings in 25 mm units.
- Anvil retractor device for repeated measurements.
- Adjustable work- table.
- Linear friction free motion transfer mechanism between anvil and indicator for high accuracy.
- Anvil pressure adjustable and remains constant.

Groove Micrometers

These micrometers are designed for measuring grooves, recesses and shoulders located inside a bore. They have standard (12.7 mm) dia discs. 6.35 mm dia discs are used to reach hard to get at locations inside a small bore.

All disc thickness are 0.75 mm and are hardened and lapped to minimize parallax and to achieve a higher degree of accuracy.

These Groove Micrometers measure not only thickness and spacing of grooves, but also measure from an edge to a land, or from shoulder to groove. Micrometers are satin-chrome finished throughout.

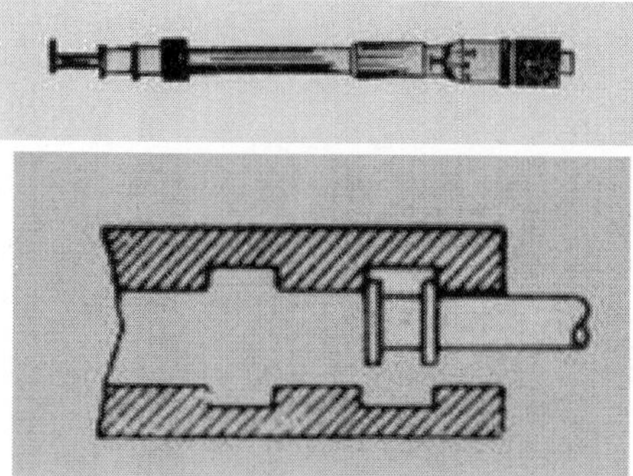

Figure: *Outside measure ment*

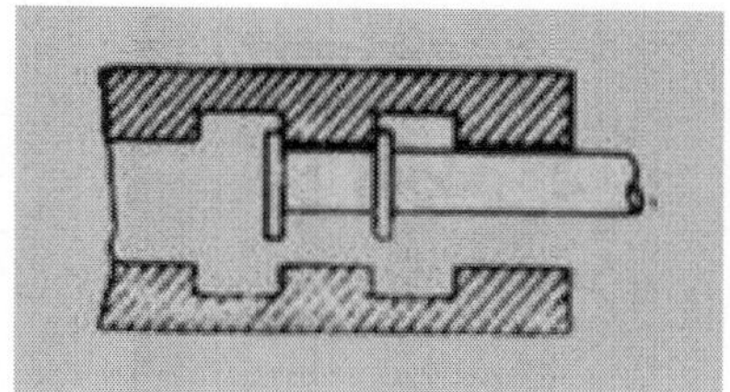

Figure: *inside measurement (add 1.5 mm to Reading)*

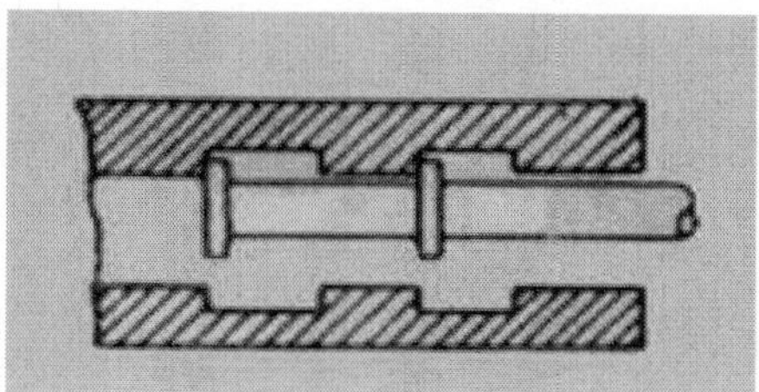

Figure: *Edge to Edge measurement*

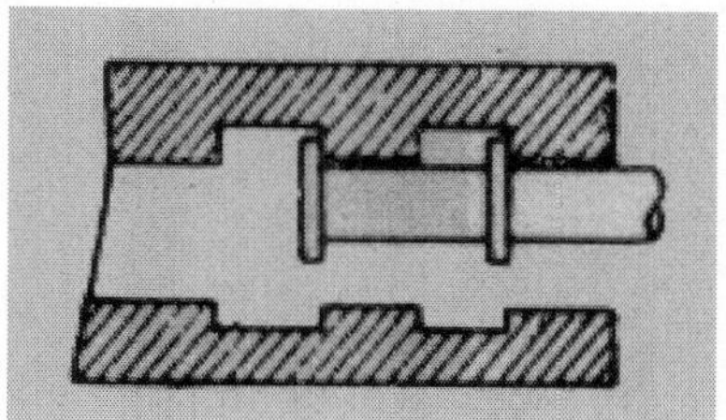

Figure: *Edge to Edge measurement (add 0.75 mm to Reading) (add 0.75 mm to Reading)*

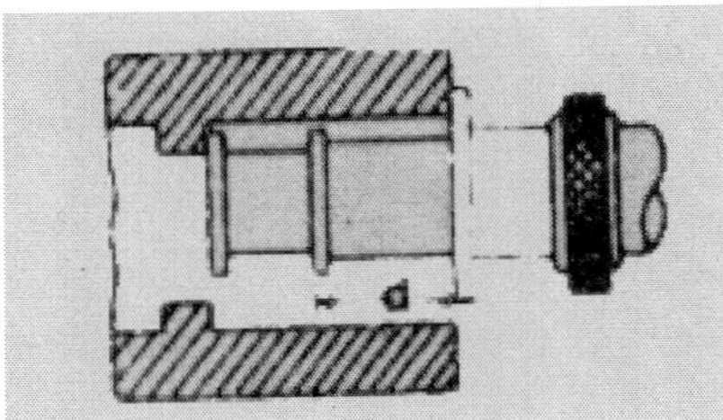

Figure: *Edge to Edge measurement with Locating disc. (add 1.5 mm to Reading)*

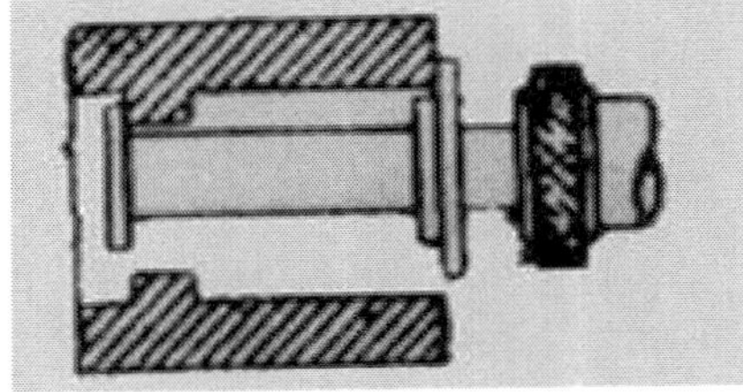

Figure: *Edge to Edge measurement with Locating disc. (add 0.75 mm to Reading)*

Digital Micrometers

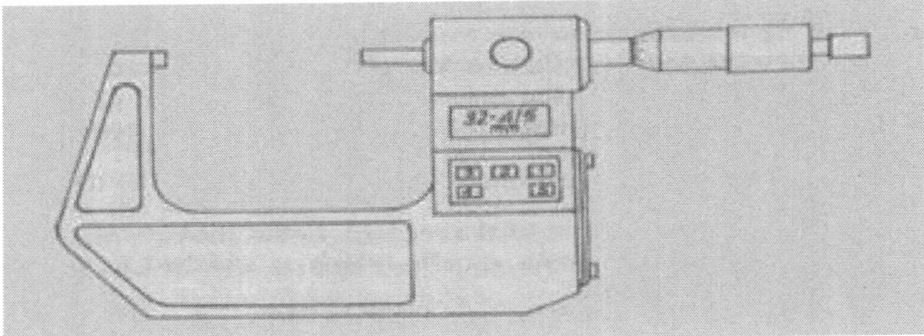

Special Features:

1- Stainless steel spindles.
2- Every hundredth in metric (every thousandth In inches) is numbered for easy reading.
3- Measuring faces carbide tipped for a long tool life.
4- Friction thimble or ratchet stop for exact and repetitive readings
5- Spindle thread hardened, ground and lapped.
6- Positive locking clamp ensures locking of spindle at any desired setting.
7- Large diameter pearl chrome thimbles for easy reading.
8- Clear, crisp black graduations and figures for easy reading against stain-chrome finish.

9- Anvil and spindle hardened and precision ground with micro-lap finish on ends.

10- Quick and easy wear adjustment.

11- Operation is simple with push button controls for "Zero" reset and indication "hold".

Specifications:

Thimble graduation: 0.01 mm

Spindle screw thread: 0.5 mm pitch

Feed error: ± 0.001 mm (20° C)

Quantizing error: ± 1 count max

Resolution: 0.001 mm

Ranges: 25, 50, 75 and 100 mm

Measuring force: 6 ~ 10 N (constant)

Ambient temperature: 0 ~ 40° C (operation) ÜÜÜ 10 ~ 60° C (storage)

Weight: 430 gm.

Control Push Buttons

1- ON/OFF: power ON or OFF.

2- IN/MM: to select the unit (in or mm) of measurement.

At power on, the unit of mm is present.

3- ZERO: to reset the display to zero at any desired position.

4- HOLD: to hold measurements taken until you push the button again. Then, the display shows up the current position of the spindle measured from predetermined zero point. The "hold" condition can also be released by the ZERO button.

5- ORIGIN: for LCD Micrometers range 25-50 mm (1-2") and above, ORIGIN button sets minimum value for the micrometer depending on its size. Micrometer count then commences from this value.

Alarm indication showing excessive speed of thimble is provided and alarm can be re-set by the ZERO buttons.

Compass

A compass is a navigational instrument that shows directions in a frame of reference that is stationary relative to the surface of the earth. The frame of reference defines the four *cardinal directions* (or *points*) – north, south, east, and west. Intermediate directions are also defined. Usually, a diagram called a compass rose, which shows the directions (with their names usually abbreviated to initials), is marked on the compass. When the compass is in use, the rose is aligned with the real directions in the frame of reference, so, for example, the "N" mark on the rose really points to the north.

Frequently, in addition to the rose or sometimes instead of it, angle markings in degrees are shown on the compass. North corresponds to zero degrees, and the angles increase clockwise, so east is 90 degrees, south is 180, and west is 270. These numbers allow the compass to show azimuths or bearings, which are commonly stated in this notation.

The magnetic compass was first invented as a device for divination as early as the ChineseHan Dynasty (since about 206 BC). The compass was used in Song Dynasty China by the military for navigational orienteering by 1040-1044, and was used for maritime navigation by 1111 to 1117. The use of a compass is recorded in Western Europe between 1187 and 1202, and in Persia in 1232. The dry compass was invented in Europe around 1300. This was supplanted in the early 20th century by the liquid-filled magnetic compass.

Types of Compasses

There are two widely used and radically different types of compass. The magnetic compass contains a magnet that interacts with the earth's magnetic field and aligns itself to point to the magnetic poles. Simple compasses of this type show directions in a frame of reference in which the directions of the magnetic poles are due north and south. These directions are called *magnetic north* and *magnetic south.*

The gyro compass (sometimes spelled with a hyphen, or as one word) contains a rapidly spinning wheel whose rotation interacts dynamically with the rotation of the earth so as to make the wheel precess, losing energy to friction until its axis of rotation is parallel with the earth's.

The wheel's axis therefore points to the earth's rotational poles, and a frame of reference is used in which the directions of the rotational poles are due north and south. These directions are called *true north* and *true south,* respectively. The astrocompass works by observing the direction of stars and other celestial bodies.

Figure: *A military compass that was used during World War I.*

There are other devices which are not conventionally called compasses but which do allow the true cardinal directions to be determined. Some GPS receivers have two or three antennas, fixed some distance apart to the structure of a vehicle, usually an aircraft or ship. The exact latitudes and longitudes of the antennas can be determined simultaneously, which allows the directions of the cardinal points to be calculated relative to the heading of the aircraft (the direction in which its nose is pointing), rather than to its direction of movement, which will be different if there is a crosswind. They are said to work "like a compass", or "as a compass".

Even a GPS device or similar can be used as compass, since if the receiver is being moved, even at walking pace, it can follow the change of its position, and hence determine the compass bearing of its direction of movement, and thence the directions of the cardinal points relative to its direction of movement. A much older example was the Chinese south-pointing chariot, which worked like a compass by directional dead reckoning. It was initialized by hand, possibly using astronomical observations e.g. of the Pole Star, and thenceforth counteracted every turn that was made to keep its pointer aiming in the desired direction, usually to the south. Watches and sundials can also be used to find compass directions.

A recent development is the electronic compass which detects the direction without potentially fallible moving parts. This may use a fibre optic gyrocompass or a magnetometer. The magnetometer frequently appears as an optional subsystem built into hand-held GPS receivers and mobile phones. However, magnetic compasses remain popular, especially in remote areas, as they are relatively inexpensive, durable, and require no power supply.

Magnetic Compass

The magnetic compass consists of a magnetized pointer (usually marked on the North end) free to align itself with Earth's magnetic field. A compass is any magnetically sensitive device capable of indicating the direction of the magnetic north of a planet's magnetosphere. The face of the compass generally highlights the cardinal points of north, south, east and west. Often, compasses are built as a stand alone sealed instrument with a magnetized bar or needle turning freely upon a pivot, or moving in a fluid, thus able to point in a northerly and southerly direction.

The compass greatly improved the safety and efficiency of travel, especially ocean travel. A compass can be used to calculate heading, used with a sextant to calculate latitude, and with a marine chronometer to calculate longitude. It thus provides a much improved navigational capability that has only been recently supplanted by modern devices such as the Global Positioning System (GPS).

How a Magnetic Compass Works

A compass functions as a pointer to "magnetic north" because the magnetized needle at its heart aligns itself with the lines of the Earth's magnetic field. The magnetic field exerts a torque on the needle, pulling one end or *pole* of the needle toward the Earth's North magnetic pole, and the other toward the South magnetic pole. The needle is mounted on a low-friction pivot point, in better compasses a jewel bearing, so it can turn easily. When the compass is held level, the needle turns until, after a few seconds to allow oscillations to die out, one end points toward the North magnetic pole.

A magnet or compass needle's "north" pole is defined as the one which is attracted to the North magnetic pole of the Earth, in northern Canada. Since opposite poles attract ("north" to "south") the North magnetic pole of the Earth is actually the *south* pole of the Earth's magnetic field. The compass needle's north pole is always marked in some way: with a distinctive colour, luminous paint, or an arrowhead.

Figure: *An inexpensive compass, aligned so that its needle points through the "North" mark on its compass card.*

Instead of a needle, professional compasses usually have bar magnets glued to the underside of a disk pivoted in the centre so it can turn, called a "compass card", with a "compass rose" showing the cardinal points and degrees marked on it. Better compasses are "*liquid-filled*"; the chamber containing the needle or disk is filled with a liquid whose purpose is to damp the oscillations of the needle so it will settle down to point to North more quickly, and also to protect the needle or disk from shock.

In navigation, directions on maps are expressed with reference to *geographical* or *true north*, the direction toward the Geographical North Pole, the rotation axis of the Earth. Since the Earth's magnetic poles are near, but are not at the same locations as its geographic poles, a compass does not point to true north. The direction a compass points is called *magnetic north*, the direction of the North magnetic pole, located in northeastern Canada. Depending on where the compass is located on the surface of the Earth the angle between true north and magnetic north, called *magnetic declination* can vary widely, increasing the farther one is from the prime meridian of the Earth's magnetic field. The local magnetic declination is given on most maps, to allow the map to be oriented with a compass parallel to true north. Some magnetic compasses include means to manually compensate for the magnetic declination, so that the compass shows true directions.

In geographic regions near the magnetic poles, in northeastern Canada and Antarctica, variations in the Earth's magnetic field cause magnetic compasses to have such large errors that they are useless, so other instruments must be used for navigation. The positions of the magnetic poles change over time on a time-scale that is not extremely long by human standards. Significant movements happen in a few years.

History

The compass was invented in China, during the Han Dynasty between the 2nd century BC and 1st century AD. The first compasses were made of lodestone, a naturally magnetized ore of iron. Ancient Chinese people found that if a lodestone was suspended so it could turn freely, it would always point in the same direction, toward the magnetic poles. Early compasses were used for geomancy "in the search for gems and the selection of sites for houses," but were later adapted for navigation during the Song Dynasty in the 11th century. Later compasses were made of iron needles, magnetized by striking them with a lodestone. The dry compass was invented in medieval Europe around 1300. This was supplanted in the early 20th century by the liquid-filled magnetic compass.

Navigation Prior to the Compass

Prior to the introduction of the compass, position, destination, and direction at sea were primarily determined by the sighting of landmarks, supplemented with the observation of the position of celestial bodies. On cloudy days, the Vikings may have used cordierite or some other birefringent crystal to determine the sun's direction and elevation from the polarization of daylight; their astronomical knowledge was sufficient to let them use this information to determine their proper heading.

For more southerly Europeans unacquainted with this technique, the invention of the compass enabled the determination of heading when the sky was overcast or foggy. This enabled mariners to navigate safely far from land, increasing sea trade, and contributing to the Age of Discovery.

Geomancy and Feng Shui

Magnetism was originally used, not for navigation, but for geomancy and fortune-telling by the Chinese. The earliest Chinesemagnetic compasses were probably not designed for navigation, but rather to order and harmonize their environments and buildings in accordance with the geomantic principles of *feng shui*. These early compasses were made using lodestone, a special form of the mineral magnetite that aligns itself with the Earth's magnetic field.

Based on Krotser and Coe's discovery of an Olmechematite artifact in Mesoamerica, radiocarbon dated to 1400-1000 BC, astronomer John Carlson has hypothesized that the Olmec might have used the geomagnetic lodestone earlier than 1000 BC for geomancy, a method of divination, which if proven true, predates the Chinese use of magnetism for feng shui by a millennium. Carlson speculates that the Olmecs used similar artifacts as a directional device for astronomical or geomantic purposes but does not suggest navigational usage. The artifact is part of a polished hematite (lodestone) bar with a groove at one end (possibly for sighting). The artifact now consistently points 35.5 degrees west of north, but may have pointed north-south when whole. Carlson's claims have been disputed by other scientific researchers, who have suggested that the artifact is actually a constituent piece of a decorative ornament and not a purposely built compass. Several other hematite or magnetite artifacts have been found at pre-Columbian archaeological sites in Mexico and Guatemala.

Navigational Compass

The invention of the navigational compass is credited by scholars to the Chinese, who began using it for navigation sometime between the 9th and 11th century, "some time before 1050, possibly as early as 850." A common theory by historians, suggests that the Arabs introduced the compass from China to Europe, although current textual evidence only supports the fact that Chinese use of the navigational compass preceded that of Europe and the Middle East.

China

There is disagreement as to exactly when the compass was invented. These are noteworthy Chinese literary references in evidence for its antiquity:

- The earliest Chinese literature reference to magnetism lies in the 4th century BC writings of Wang Xu: "The lodestone attracts iron." The book also notes that the people of the state of Zheng always knew their position by means of a "south-pointer"; some authors suggest that this refers to early use of the compass.

Figure: *Model of a Han Dynasty (206 BC–220 AD) south-indicating ladle or* sinan. *It is theorized that the south-pointing spoons of the Han dynasty were magnetized lodestones.*

- The first mention of a spoon, speculated to be a lodestone, observed pointing in a cardinal direction is a Chinese work composed between 70 and 80 AD (*Lunheng*), which records that "But when the south pointing spoon is thrown upon the ground, it comes to rest pointing at the south." Within the text, the author Wang Chong describes the spoon as a phenomenon that he has personally observed. Although the passage does not explicitly mention magnetism, according to Chen-Cheng Yih, the "device described by Wang Chong has been widely considered to be the earliest form of the magnetic compass."
- The first clear account of magnetic declination occurs in the *Kuan Shih Ti Li Chih Meng* ("Mr. Kuan's Geomantic Instructor"), dating to 880. Another text, the *Chiu Thien Hsuan Nu Chhing Nang Hai Chio Ching* ("Blue Bag Sea Angle Manual") from around the same period, also has an implicit description of magnetic declination. It has been argued that this knowledge of declination requires the use of the compass.

- A reference to a magnetized needle as a "mysterious needle" appears in 923-926 in the *Chung Hua Ku Chin Chu* text written by Ma Kao. The same passage is also attributed to the 4th century AD writer Tshui Pao, although it is postulated that the former text is more authentic. The shape of the needle is compared to that of a tadpole, and may indicate the transition between "lodestone spoons" and "iron needles."
- The earliest reference to a specific magnetic direction finder device for land navigation is recorded in a Song Dynasty book dated to 1040-44. There is a description of an iron "south-pointing fish" floating in a bowl of water, aligning itself to the south. The device is recommended as a means of orientation "in the obscurity of the night."

  The *Wujing Zongyao* "Collection of the Most Important Military Techniques") stated: "When troops encountered gloomy weather or dark nights, and the directions of space could not be distinguished...they made use of the [mechanical] south-pointing carriage, or the south-pointing fish." This was achieved by heating of metal (especially if steel), known today as thermoremanence, and would have been capable of producing a weak state of magnetization. While the Chinese achieved magnetic remanence and induction by this time, in both Europe and Asia the phenomenon was attributed to the supernatural and occult, until about 1600 when William Gilbert published his *De Magnete*.
- The first incontestable reference to a magnetized needle in Chinese literature appears in 1088. The *Dream Pool Essays*, written by the Song Dynastypolymath scientist Shen Kuo, contained a detailed description of how geomancers magnetized a needle by rubbing its tip with lodestone, and hung the magnetic needle with one single strain of silk with a bit of wax attached to the centre of the needle. Shen Kuo pointed out that a needle prepared this way sometimes pointed south, sometimes north.
- The earliest explicit recorded use of a magnetic compass for maritime navigation is found in Zhu Yu's book *Pingchow Table Talks* and dates from 1111 to 1117: *The ship's pilots are acquainted with the configuration of the coasts; at night they steer by the stars , and in the daytime by the sun. In dark weather they look at the south pointing needle.*

Thus, the use of a magnetic compass by the military for land navigation occurred sometime before 1044, but incontestable evidence for the use of the compass as a maritime navigational device did not appear until 1117.

The typical Chinese navigational compass was in the form of a magnetic needle floating in a bowl of water. According to Needham, the Chinese in the Song Dynasty and continuing Yuan Dynasty did make use of a dry compass, although this type never became as widely used in China as the wet compass. Evidence of this is found in the *Shilin guangji* ("Guide Through the Forest of Affairs"), published in 1325 by Chen Yuanjing, although its compilation had taken place between 1100 and 1250.

The dry compass in China was a dry suspension compass, a wooden frame crafted in the shape of a turtle hung upside down by a board, with the lodestone sealed in by wax, and if rotated, the needle at the tail would always point in the northern cardinal direction. Although the European compass-card in box frame and dry pivot needle was adopted in China after its use was taken by Japanese pirates in the 16th century (who had in turn learned of it from Europeans), the Chinese design of the suspended dry compass persisted in use well into the 18th century. However, according to Kreutz there is only a single Chinese reference to a dry-mounted needle (built into a pivoted wooden tortoise) which is dated to between 1150 and 1250, and claims that there is no clear indication that Chinese mariners ever used anything but the floating needle in a bowl until the 16th-century.

The first recorded use of a 48 position mariner's compass on sea navigation was noted in *The Customs of Cambodia* by Yuan Dynasty diplomat Zhou Daguan, he described his 1296 voyage from Wenzhou to Angkor Thom in detail; when his ship set sail from Wenzhou, the mariner took a needle direction of "ding wei" position, which is equivalent to 22.5 degree SW. After they arrived at Baria, the mariner took "Kun Shen needle", or 52.5 degree SW. Zheng He's Navigation Map, also known as "The Mao Kun Map", contains a large amount of detail "needle records" of Zheng He's expeditions.

There is a debate over the diffusion of the compass after its first appearance with the Chinese. At present, according to Kreutz, scholarly consensus is that the Chinese invention predates the first European mention by 150 years.

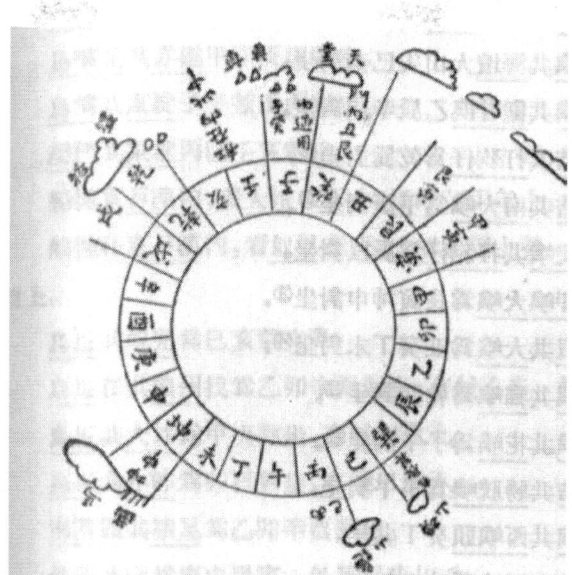

Figure: *Diagram of a Ming Dynasty mariner's compass*

However, there are questions over diffusion, because of the apparent failure of the Arabs to function as possible intermediaries between East and West because of the earlier recorded appearance of the compass in Europe (1190) than in the Muslim world (1232, 1242, and 1282). The first European mention of a magnetized needle and its use among sailors occurs in Alexander Neckam's *De naturis rerum* (On the Natures of Things), written in 1190.

The earliest reference to a compass in the Middle East is attributed to the Persians, who describe an iron fish-like compass in a talebook dating from 1232. In the Arab world, the earliest reference comes in *The Book of the Merchants' Treasure,* written by one Baylak al-Kibjaki in Cairo about 1282. Since the author describes having witnessed the use of a compass on a ship trip some forty years earlier, some scholars are inclined to antedate its first appearance accordingly. That the Arabic word for "Compass" (*al-konbas*) may be a derivation of the old Italian word for compass, is also used as evidence for the lack of diffusion from China to Europe. However, the Persian compass is described as fish-like, which is a characteristic of early Chinese compasses from the 11th century, suggesting transmission from China to Persia.

Medieval Europe

Alexander Neckam reported the use of a magnetic compass for the region of the English Channel in the texts *De utensilibus* and *De naturis rerum*, written between 1187 and 1202, after he returned to England from France and prior to entering the Augustinian abbey at Cirencester. In 1269 Petrus Peregrinus of Maricourt described a floating compass for astronomical purposes as well as a dry compass for seafaring, in his well-known Epistola de magnete.

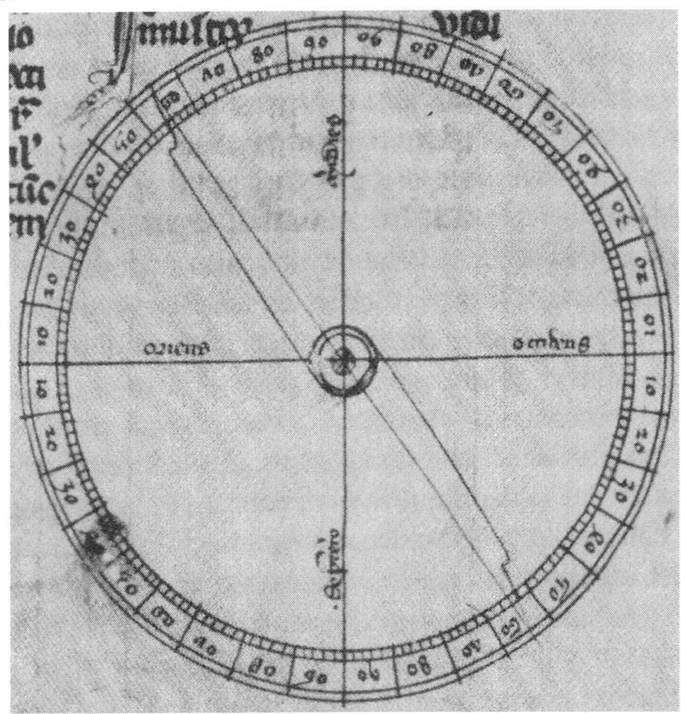

Figure: *Pivoting compass needle in a 14th-century copy of* Epistola de magnete *of Peter Peregrinus (1269).*

In the Mediterranean, the introduction of the compass, at first only known as a magnetized pointer floating in a bowl of water, went hand in hand with improvements in dead reckoning methods, and the development of Portolan charts, leading to more navigation during winter months in the second half of the 13th century.

While the practice from ancient times had been to curtail sea travel between October and April, due in part to the lack of dependable clear skies during the Mediterranean winter, the prolongation of the sailing season resulted in a gradual, but sustained increase in shipping movement; by around 1290 the sailing season could start in late January or February, and end in December. The additional few months were of considerable economic importance. For instance, it enabled Venetian convoys to make two round trips a year to the Levant, instead of one. At the same time, traffic between the Mediterranean and northern Europe also increased, with first evidence of direct commercial voyages from the Mediterranean into the English Channel coming in the closing decades of the 13th century, and one factor may be that the compass made traversal of the Bay of Biscay safer and easier. However, critics like Kreutz feel that it was later in 1410 that anyone really started steering by compass.

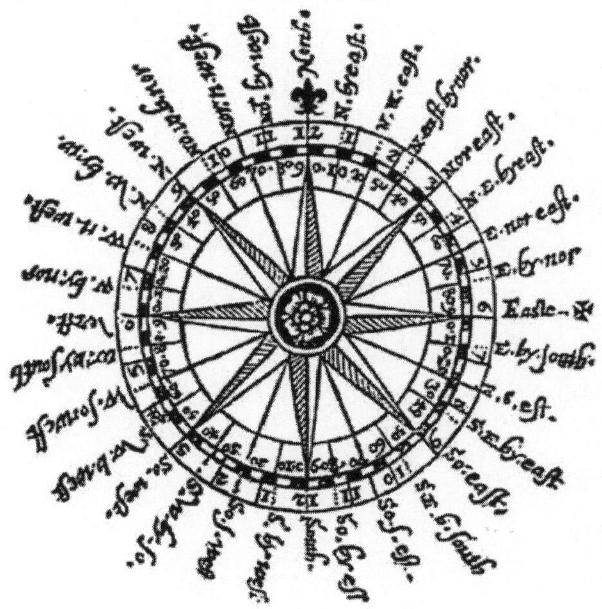

Figure: *Navigational sailor's compass rose.*

At present, according to Kreutz, "barring the discovery of new evidence, it seems clear the first Chinese reference to" the compass "antedates any European mention by roughly 150 years." However, there are questions over diffusion, because of the apparent failure of the Arabs to function as possible intermediaries between East and West because of the earlier recorded appearance of the compass in Europe (1190) than in the Muslim world (1232, 1242, and 1282). This is countered by evidence of the temporal proximity of the Chinese navigational compass (1117) to its first appearance in Europe (1190) and the common shape of the early compass as a magnetized needle floating in a bowl of water.

Islamic World

The earliest reference to an iron fish-like compass in the Islamic world occurs in a Persian talebook from 1232. This fish shape was from a typical early Chinese design. The earliest Arabic reference to a compass — in the form of magnetic needle in a bowl of water — comes from the Yemeni sultan and astronomer Al-Ashraf in 1282. He also appears to be the first to make use of the compass for astronomical purposes. Since the author describes having witnessed the use of a compass on a ship trip some forty years earlier, some scholars are inclined to antedate its first appearance in the Arab world accordingly.

India

The compass was used in India for navigational purposes and was known as the matsya yantra, because of the placement of a metallic fish in a cup of oil.

Medieval Africa

There is evidence that the distribution of the compass from China likely also reached eastern Africa by way of trade through the end of the Silk Road that ended in East African centre of trade in Somalia and the Swahili city-state kingdoms. There is evidence that Swahili maritime merchants and sailors acquired the compass at some point and used them for navigation of Swahili versions of dhows.

Later Developments

Dry Compass

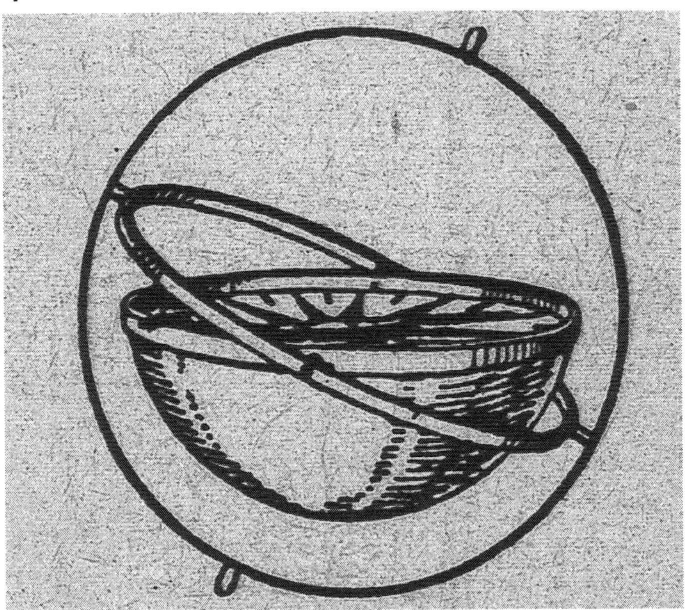

Figure: *Early modern dry compass suspended by a gimbal (1570).*

The dry mariner's compass was invented in Europe around 1300. The dry mariner's compass consists of three elements: A freely pivoting needle on a pin enclosed in a little box with a glass cover and a wind rose, whereby "the wind rose or compass card is attached to a magnetized needle in such a manner that when placed on a pivot in a box fastened in line with the keel of the ship the card would turn as the ship changed direction, indicating always what course the ship was on".

Later, compasses were often fitted into a gimbal mounting to reduce grounding of the needle or card when used on the pitching and rolling deck of a ship.

While pivoting needles in glass boxes had already been described by the French scholar Peter Peregrinus in 1269, and by the Egyptian scholar Ibn Sim¿ûn in 1300, traditionally Flavio Gioja (fl. 1302), an Italian pilot from Amalfi, has been credited with perfecting the sailor's compass by suspending its needle over a compass card, thus giving the compass its familiar appearance. Such a compass with the needle attached to a rotating card is also described in a commentary on Dante's *Divine Comedy* from 1380, while an earlier source refers to a portable compass in a box (1318), supporting the notion that the dry compass was known in Europe by then.

Bearing Compass

A *bearing compass* is a magnetic compass mounted in such a way that it allows the taking of bearings of objects by aligning them with the lubber line of the bearing compass. A *surveyor's compass* is a specialized compass made to accurately measure heading of landmarks and measure horizontal angles to help with map making.

These were already in common use by the early 18th century and are described in the 1728 Cyclopaedia. The bearing compass was steadily reduced in size and weight to increase portability, resulting in a model that could be carried and operated in one hand. In 1885, a patent was granted for a hand compass fitted with a viewing prism and lens that enabled the user to accurately sight the heading of geographical landmarks, thus creating the *prismatic compass*. Another sighting method was by means of a reflective mirror. First patented in 1902, the *Bézard compass* consisted of a field compass with a mirror mounted above it. This arrangement enabled the user to align the compass with an objective while simultaneously viewing its bearing in the mirror.

Figure: *Bearing compass (18th century).*

In 1928, Gunnar Tillander, a Swedish unemployed instrument maker and avid participant in the sport of orienteering, invented a new style of bearing compass. Dissatisfied with existing field compasses, which required a separate protractor in order to take bearings from a map, Tillander decided to incorporate both instruments into a single instrument. It combined a compass with a protractor built into the base. His design featured a metal compass capsule containing a magnetic needle with orienting marks mounted into a transparent protractor baseplate with a lubber line (later called a *direction of travel indicator*).

By rotating the capsule to align the needle with the orienting marks, the course bearing could be read at the lubber line. Moreover, by aligning the baseplate with a course drawn on a map - ignoring the needle - the compass could also function as a protractor. Tillander took his design to fellow orienteers Björn, Alvid, and Alvar Kjellström, who were selling basic compasses, and the four men modified Tillander's design. In December 1932, the Silva Company was formed with

Tillander and the three Kjellström brothers, and the company began manufacturing and selling its Silva orienteering compass to Swedish orienteers, outdoorsmen, and army officers.

Liquid Compass

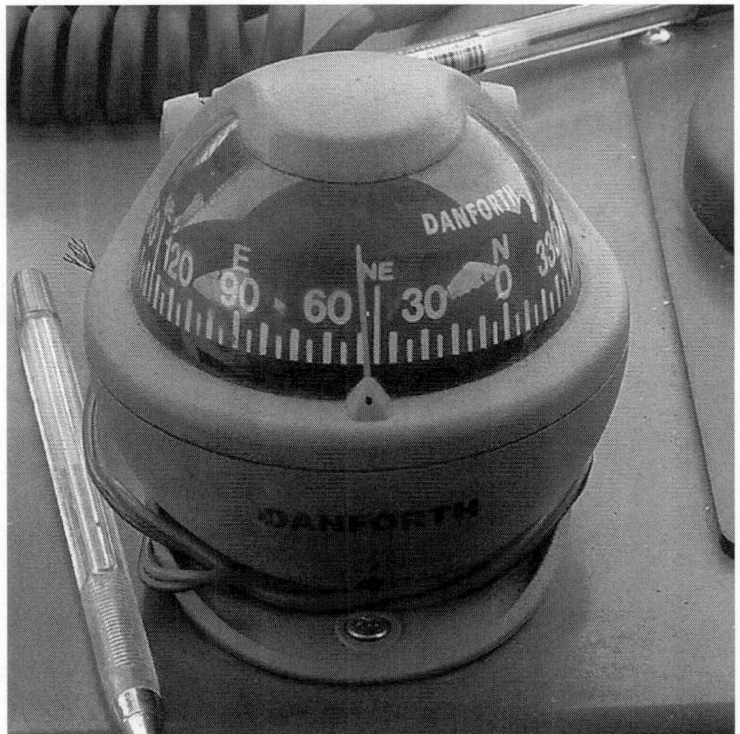

Figure: *A surface mount, liquid-filled compass on a boat.*

The liquid compass is a design in which the magnetized needle or card is damped by fluid to protect against excessive swing or wobble, improving readability while reducing wear. A rudimentary working model of a liquid compass was introduced by Sir Edmund Halley at a meeting of the Royal Society in 1690.However, as early liquid compasses were fairly cumbersome and heavy, and subject to damage, their main advantage was aboard ship. Protected in a binnacle and normally gimbal-mounted, the liquid inside the compass housing effectively damped shock and vibration, while eliminating excessive swing and grounding of the card caused by the pitch and roll of the vessel.

The first liquid mariner's compass believed practicable for limited use was patented by the Englishman Francis Crow in 1813. Liquid-damped marine compasses for ships and small boats were occasionally used by the BritishRoyal Navy from the 1830s through 1860, but the standard Admiralty compass remained a dry-mount type. In the latter year, the American physicist and inventor Edward Samuel Ritchie patented a greatly improved liquid marine compass that was adopted in revised form for general use by the United States Navy, and later purchased by the Royal Navy as well.

Despite these advances, the liquid compass was not introduced generally into the Royal Navy until 1908. An early version developed by RN Captain Creak proved to be operational under heavy gunfire and seas, but was felt to lack navigational precision compared with the design by Lord Kelvin:

Figure: *Typical aircraft-mounted magnetic compass.*

Captain Creak's first step in the development of the liquid compass was to introduce a "card mounted on a float, with two thin and relatively short needles, fitted with their poles at the scientifically correct angular distances, and with the centre of gravity, centre of buoyancy, and the point of suspension in correct relation to each other...The compass thus designed rectified the defects of the Admiralty Standard Compass...with the additional advantage of considerable steadiness under heavy gunfire and in a seaway... The one defect in the compass as developed by Creak up to 1892 was that "for manoeuvring purposes it was inferior to Lord Kelvin's compass, owing to comparative sluggishness on a large alteration of course through the drag on the card by the liquid in which it floated...

However, with ship and gun sizes continuously increasing, the advantages of the liquid compass over the Kelvin compass became unavoidably apparent to the Admiralty, and after widespread adoption by other navies, the liquid compass was generally adopted by the Royal Navy as well.

Liquid compasses were next adapted for aircraft. In 1909, Captain F.O. Creagh-Osborne, Superintendent of Compasses at the British Admiralty, introduced his *Creagh-Osborne* aircraft compass, which used a mixture of alcohol and distilled water to damp the compass card.

After the success of this invention, Capt. Creagh-Osborne adapted his design to a much smaller pocket model for individual use by officers of artillery or infantry, receiving a patent in 1915.

In December 1932, the newly founded Silva Company of Sweden introduced its first baseplate or bearing compass that used a liquid-filled capsule to damp the swing of the magnetized needle.

The liquid-damped Silva took only four seconds for its needle to settle in comparison to thirty seconds for the original version.

In 1933 Tuomas Vohlonen, a surveyor by profession, applied for a patent for a unique method of filling and sealing a lightweight celluloid compass housing or capsule with a petroleum distillate to dampen the needle and protect it from shock and wear caused by excessive motion. Introduced in a wrist-mount model in 1936 as the Suunto Oy *Model M-311*, the new capsule design led directly to the lightweight liquid field compasses of today.

History of Non-navigational Uses

Building Orientation

Evidence for the orientation of buildings by the means of a magnetic compass can be found in 12th century Denmark: one fourth of its 570 Romanesque churches are rotated by 5-15 degrees clockwise from true east-west, thus corresponding to the predominant magnetic declination of the time of their construction.

Most of these churches were built in the 12th century, indicating a fairly common usage of magnetic compasses in Europe by then.

Mining

The use of a compass as a direction finder underground was pioneered by the Tuscan mining town Massa where floating magnetic needles were employed for determining tunneling and defining the claims of the various mining companies as early as the 13th century. In the second half of the 15th century, the compass became standard equipment for Tyrolian miners. Shortly afterwards the first detailed treatise dealing with the underground use of compasses was published by a German miner Rülein von Calw (1463–1525).

Astronomy

Three astronomical compasses meant for establishing the meridian were described by Peter Peregrinus in 1269 (referring to experiments made before 1248) In the 1300s, an Arabic treatise written by the Egyptian astronomer and muezzin Ibn Sim¿ûn describes a dry compass for use as a "Qibla indicator" to find the direction to Mecca. Ibn Sim¿ûn's compass, however, did not feature a compass card nor the familiar glass box.

In the 14th century, the Syrian astronomer and timekeeper Ibn al-Shatir (1304–1375) invented a timekeeping device incorporating both a universal sundial and a magnetic compass. He invented it for the purpose of finding the times of salat prayers. Arab navigators also introduced the 32-point compass rose during this time.

Modern Compasses

Modern compasses usually use a magnetized needle or dial inside a capsule completely filled with a liquid (lamp oil, mineral oil, white spirits, purified kerosene, or ethyl alcohol is common). While older designs commonly incorporated a flexible rubber diaphragm or airspace inside the capsule to allow for volume changes caused by temperature or altitude, some modern liquid compasses utilize smaller housings and/or flexible capsule materials to accomplish the same result.

The liquid inside the capsule serves to dampen the movement of the needle, reducing oscillation time and increasing stability. Key points on the compass, including the north end of the needle are often marked with phosphorescent, photoluminescent, or self-luminous materials to enable the compass to be read at night or in poor light. As the compass fill liquid is noncompressible under pressure, many ordinary liquid-filled compasses will operate accurately underwater to considerable depths.

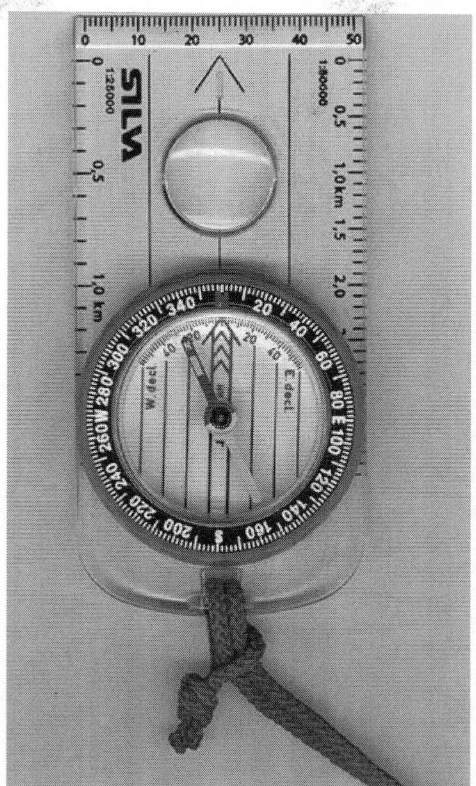

Figure: *A liquid-filled protractor or orienteering compass with lanyard.*

Many modern compasses incorporate a baseplate and protractor tool, and are referred to variously as "orienteering", "baseplate", "map compass" or "protractor" designs. This type of compass uses a separate magnetized needle inside a rotating capsule, an orienting "box" or gate for aligning the needle with magnetic north, a transparent base containing map orienting lines, and a bezel (outer dial) marked in degrees or other units of angular measurement. The capsule is mounted in a transparent baseplate containing a *direction-of-travel* (DOT) indicator for use in taking bearings directly from a map.

Other features found on modern orienteering compasses are map and romer scales for measuring distances and plotting positions on maps, luminous markings on the face or bezels, various sighting mechanisms (mirror, prism, etc.) for taking bearings of distant objects with greater precision, "global" needles for use in differing hemispheres, adjustable declination for obtaining instant true bearings without resort to arithmetic, and devices such as clinometers for measuring gradients.

The sport of orienteering has also resulted in the development of models with extremely fast-settling and stable needles for optimal use with a topographic map, a land navigation technique known as *terrain association.*

Figure: *Liquid filled lensatic compass.*

The military forces of a few nations, notably the United States Army, continue to issue field compasses with magnetized compass dials or cards instead of needles. A magnetic card compass is usually equipped with an optical, lensatic, or prismatic sight, which allows the user to read the bearing or azimuth off the compass card while simultaneously aligning the compass with the objective. Magnetic card compass designs normally require a separate protractor tool in order to take bearings directly from a map.

The U.S. M-1950 military lensatic compass does not use a liquid-filled capsule as a dampening mechanism, but rather electromagnetic induction to control oscillation of it magnetized card. A "deep-well" design is used to allow the compass to be used globally with a card tilt of up to 8 degrees without impairing accuracy. As induction forces provide less damping than liquid-filled designs, a needle lock is fitted to the compass to reduce wear, operated by the folding action of the rear sight/lens holder. The use of air-filled induction compasses has declined over the years, as they may become inoperative or inaccurate in freezing temperatures or extremely humid environments due to condensation or water ingress.

Figure: *Cammenga air filled lensatic compass.*

Some military compasses, like the U.S. M-1950 (Cammenga 3H) military lensatic compass, the Silva 4b *Militaire*, and the Suunto M-5N(T) contain the radioactive material tritium ($_1H^3$) and a combination of phosphors. The U.S. M-1950 equipped with self-luminous lighting contains 120 mCi (millicuries) of tritium. The purpose of the tritium and phosphors is to provide illumination for the compass, via radioluminescenttritium illumination, which does not require the compass to be "recharged" by sunlight or artificial light. However, tritium has a half-life of only about 12 years, so a compass that contains 120 mCi of tritium when new will contain only 60 when it is 12 years old, 30 when it is 24 years old, and so on. Consequently, the illumination of the display will fade.

Mariner's compasses can have two or more gimbaled magnets permanently attached to a compass card. These move freely on a pivot. A *lubber line*, which can be a marking on the compass bowl or a small fixed needle indicates the ship's heading on the compass card.

Traditionally the card is divided into thirty-two points (known as *rhumbs*), although modern compasses are marked in degrees rather than cardinal points. The glass-covered box (or bowl) contains a suspended gimbal within a binnacle. This preserves the horizontal position.

Thumb Compass

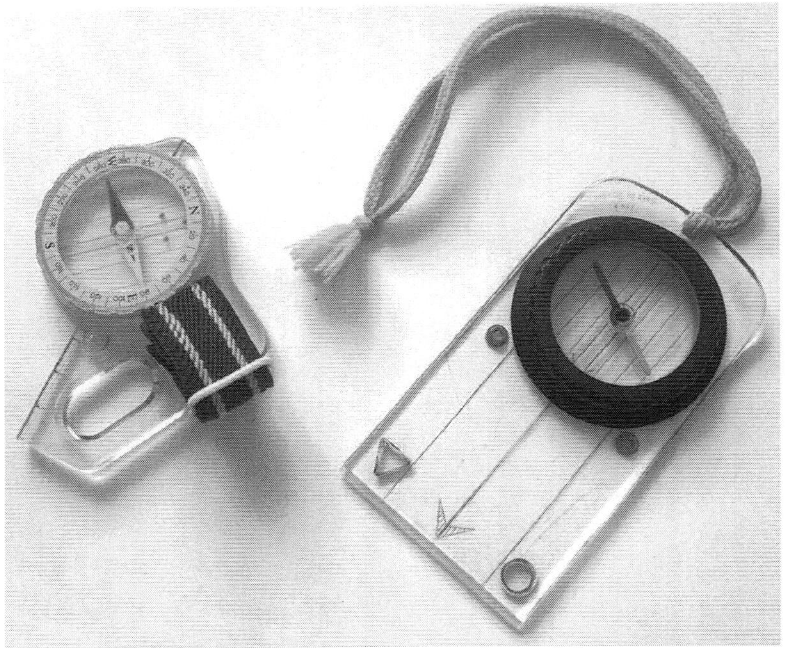

Figure: *Thumb compass on left.*

A thumb compass is a type of compass commonly used in orienteering, a sport in which map reading and terrain association are paramount. Consequently, most thumb compasses have minimal or no degree markings at all, and are normally used only to orient the map to magnetic north. Thumb compasses are also often transparent so that an orienteer can hold a map in the hand with the compass and see the map through the compass.

Gyrocompass

A *gyrocompass* is similar to a gyroscope. It is a non-magnetic compass that finds true north by using an (electrically powered) fast-spinning wheel and friction forces in order to exploit the rotation of the Earth. Gyrocompasses are widely used on ships. They have two main advantages over magnetic compasses:

- they find *true north*, i.e., the direction of Earth's rotational axis, as opposed to magnetic north,
- they are not affected by ferromagnetic metal (including iron, steel, cobalt, nickel, and various alloys) in a ship's hull. (No compass is affected by nonferromagnetic metal, although a magnetic compass will be affected by any kind of wires with electric current passing through them.)

Large ships typically rely on a gyrocompass, using the magnetic compass only as a backup. Increasingly, electronic fluxgate compasses are used on smaller vessels. However, magnetic compasses are still widely in use as they can be small, use simple reliable technology, are comparatively cheap, often easier to use than GPS, require no energy supply, and unlike GPS, are not affected by objects, e.g. trees, that can block the reception of electronic signals.

Solid State Compasses

Small compasses found in clocks, mobile phones, and other electronic devices are solid-state compasses, usually built out of two or three magnetic field sensors that provide data for a microprocessor. The correct heading relative to the compass is calculated using trigonometry.

Often, the device is a discrete component which outputs either a digital or analog signal proportional to its orientation. This signal is interpreted by a controller or microprocessor and used either internally, or sent to a display unit. The sensor uses highly calibrated internal electronics to measure the response of the device to the Earth's magnetic field.

GPS receivers using two or more antennae can now achieve 0.5° in heading accuracy and have startup times in seconds rather than hours for gyrocompass systems. Manufactured primarily for maritime applications, they can also detect pitch and roll of ships.

Speciality Compasses

Apart from navigational compasses, other speciality compasses have also been designed to accommodate specific uses. These include:

- Qibla compass, which is used by Muslims to show the direction to Mecca for prayers.
- Optical or prismatichand-bearing compass, most often used by surveyors, but also by cave explorers, foresters, and geologists.

This compasses ordinarily uses a liquid-damped capsule and magnetized floating compass dial with an integral optical (direct or lensatic) or prismatic sight, often fitted with built-in photoluminescent or battery-powered illumination. Using the optical or prism sight, such compasses can be read with extreme accuracy when taking bearings to an object, often to fractions of a degree. Most of these compasses are designed for heavy-duty use, with high-quality needles and jeweled bearings, and many are fitted for tripod mounting for additional accuracy.

- Trough compasses, mounted in a rectangular box whose length was often several times its width, date back several centuries. They were used for land surveying, particularly with plane tables.

Figure: *A standard Brunton Geo, used commonly by geologists.*

Limitations of the Magnetic Compass

The compass is very stable in areas close to the equator, which is far from "magnetic north".

As the compass is moved closer and closer to one of the magnetic poles of the Earth, the compass becomes more sensitive to crossing its magnetic field lines. At some point close to the magnetic pole the compass will not indicate any particular direction but will begin to drift. Also, the needle starts to point up or down when getting closer to the poles, because of the so-called magnetic inclination. Cheap compasses with bad bearings may get stuck because of this and therefore indicate a wrong direction.

All magnetic devices are subject to fields other than Earth's, which is not particularly strong. Local environments may contain mineral deposits and human sources such as MRIs.

Vehicles may contain ferrous metals, which may pick up their own fields. Cars may be mostly steel, and render simple compasses useless after time. While ships, submarines, and spacecraft may be built from carefully controlled materials, and later degaussed, drivers rarely take such a step.

A compass is also subject to errors when the compass is accelerated or decelerated in an airplane or automobile. Depending on which of the Earth's hemispheres the compass is located and if the force is acceleration or deceleration the compass will increase the indicated heading or decrease the indicated heading.

Another error of the mechanical compass is turning error. When one turns from a heading of east or west the compass will lag behind the turn or lead ahead of the turn. Magnetometers, and substitutes such as gyrocompasses, are more stable in such situations.

Construction of a Compass

Magnetic Needle

A magnetic rod is required when constructing a compass. This can be created by aligning an iron or steel rod with Earth's magnetic field and then tempering or striking it.

However, this method produces only a weak magnet so other methods are preferred. For example, a magnetised rod can be created by repeatedly rubbing an iron rod with a magnetic lodestone.

This magnetised rod (or magnetic needle) is then placed on a low friction surface to allow it to freely pivot to align itself with the magnetic field. It is then labelled so the user can distinguish the north-pointing from the south-pointing end; in modern convention the north end is typically marked in some way.

Needle-and-bowl Device

If a needle is rubbed on a lodestone or other magnet, the needle becomes magnetized. When it is inserted in a cork or piece of wood, and placed in a bowl of water it becomes a compass. Such devices were universally used as compass until the invention of the box-like compass with a 'dry' pivoting needle sometime around 1300.

Points of the Compass

Figure: *Wrist compass of the Soviet Army with counterclockwise double graduation: 60° (like a watch) and 360°.*

Originally, many compasses were marked only as to the direction of magnetic north, or to the four cardinal points (north, south, east, west). Later, these were divided, in China into 24, and in Europe into 32 equally spaced points around the compass card.

In the modern era, the 360-degree system took hold. This system is still in use today for civilian navigators. The degree system spaces 360 equidistant points located clockwise around the compass dial. In the 19th century some European nations adopted the "grad" (also called grade or gon) system instead, where a right angle is 100 grads to give a circle of 400 grads. Dividing grads into tenths to give a circle of 4000 decigrades has also been used in armies.

Most military forces have adopted the French "millieme" system. This is an approximation of a milli-radian (6283 per circle), in which the compass dial is spaced into 6400 units or "mils" for additional precision when measuring angles, laying artillery, etc. The value to the military is that one angular mil subtends approximately one metre at a distance of one kilometer. Imperial Russia used a system derived by dividing the circumference of a circle into chords of the same length as the radius. Each of these was divided into 100 spaces, giving a circle of 600.

The Soviet Union divided these into tenths to give a circle of 6000 units, usually translated as "mils". This system was adopted by the former Warsaw Pact countries (Soviet Union, GDR etc.), often counterclockwise. This is still in use in Russia.

Compass Balancing (Magnetic Dip)

Because the Earth's magnetic field's inclination and intensity vary at different latitudes, compasses are often balanced during manufacture so that the dial or needle will be level, eliminating needle drag which can give inaccurate readings. Most manufacturers balance their compass needles for one of five zones, ranging from zone 1, covering most of the Northern Hemisphere, to zone 5 covering Australia and the southern oceans. This individual zone balancing prevents excessive dipping of one end of the needle which can cause the compass card to stick and give false readings.

Some compasses feature a special needle balancing system that will accurately indicate magnetic north regardless of the particular magnetic zone. Other magnetic compasses have a small sliding counterweight installed on the needle itself. This sliding counterweight, called a 'rider', can be used for counterbalancing the needle against the dip caused by inclination if the compass is taken to a zone with a higher or lower dip.

Compass Correction

Like any magnetic device, compasses are affected by nearby ferrous materials, as well as by strong local electromagnetic forces. Compasses used for wilderness land navigation should not be used in proximity to ferrous metal objects or electromagnetic fields (car electrical systems, automobile engines, steel pitons, etc.) as that can affect their accuracy. Compasses are particularly difficult to use accurately in or near trucks, cars or other mechanized vehicles even when corrected for deviation by the use of built-in magnets or other devices. Large amounts of ferrous metal combined with the on-and-off electrical fields caused by the vehicle's ignition and charging systems generally result in significant compass errors.

At sea, a ship's compass must also be corrected for errors, called deviation, caused by iron and steel in its structure and equipment. The ship is *swung*, that is rotated about a fixed point while its heading is noted by alignment with fixed points on the shore. A compass deviation card is prepared so that the navigator can convert between compass and magnetic headings. The compass can be corrected in three ways. First the lubber line can be adjusted so that it is aligned with the direction in which the ship travels, then the effects of permanent magnets can be corrected for by small magnets fitted within the case of the compass. The effect of ferromagnetic materials in the compass's environment can be corrected by two iron balls mounted on either side of the compass binnacle. The coefficient a_0 representing the error in the lubber line, while the a_1, b_1 ferromagnetic effects and a_2, b_2 the non-ferromagnetic component.

Figure: *A binnacle containing a ship's standard compass, with the two iron balls which correct the effects of ferromagnetic materials. This unit is on display in a museum.*

A similar process is used to calibrate the compass in light general aviation aircraft, with the compass deviation card often mounted permanently just above or below the magnetic compass on the instrument panel. Fluxgate electronic compasses can be calibrated automatically, and can also be programmed with the correct local compass variation so as to indicate the true heading.

Using a Compass

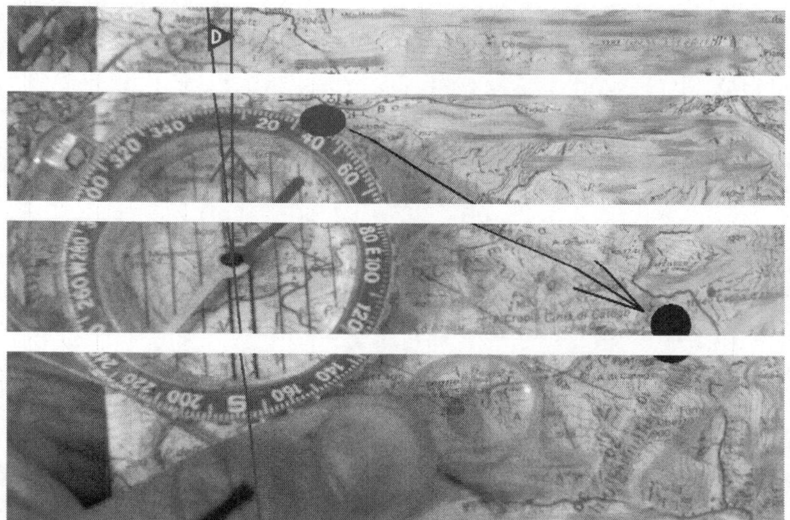

Figure: *Turning the compass scale on the map (D - the local magnetic declination).*

Figure: *When the needle is aligned with and superimposed over the outlined orienting arrow on the bottom of the capsule, the degree figure on the compass ring at the direction-of-travel (DOT) indicator gives the magnetic bearing to the target (mountain).*

A magnetic compass points to magnetic north pole, which is approximately 1,000 miles from the true geographic North Pole. A magnetic compass's user can determine true North by finding the magnetic north and then correcting for variation and deviation. Variation is defined as the angle between the direction of true (geographic) north and the direction of the meridian between the magnetic poles. Variation values for most of the oceans had been calculated and published by 1914. Deviation refers to the response of the compass to local magnetic fields caused by the presence of iron and electric currents; one can partly compensate for these by careful location of the compass and the placement of compensating magnets under the compass itself. Mariners have long known that these measures do not completely cancel deviation; hence, they performed an additional step by measuring the compass bearing of a landmark with a known magnetic bearing. They then pointed their ship to the next compass point and measured again, graphing their results. In this way, correction tables could be created, which would be consulted when compasses were used when travelling in those locations.

Mariners are concerned about very accurate measurements; however, casual users need not be concerned with differences between magnetic and true North. Except in areas of extreme magnetic declination variance (20 degrees or more), this is enough to protect from walking in a substantially different direction than expected over short distances, provided the terrain is fairly flat and visibility is not impaired. By carefully recording distances (time or paces) and magnetic bearings travelled, one can plot a course and return to one's starting point using the compass alone.

Compass navigation in conjunction with a map (*terrain association*) requires a different method. To take a map bearing or *true bearing* (a bearing taken in reference to true, not magnetic north) to a destination with a protractor compass, the edge of the compass is placed on the map so that it connects the current location with the desired destination (some sources recommend physically drawing a line). The orienting lines in the base of the compass dial are then rotated to align with actual or true north by aligning them with a marked line of longitude (or the vertical margin of the map), ignoring the compass needle entirely. The resulting *true bearing* or map bearing may then be read at the degree indicator or direction-of-travel (DOT) line, which may be followed as an *azimuth* (course) to the destination. If a *magnetic* north bearing or *compass bearing* is desired, the compass must be adjusted by the amount of magnetic declination before using the bearing so that both map and compass are in agreement. In the given example, the large mountain in the second photo was selected as the target destination on the map. Some compasses allow the scale to be adjusted to compensate for the local magnetic declination; if adjusted correctly, the compass will give the true bearing instead of the magnetic bearing.

The modern hand-held protractor compass always has an additional direction-of-travel (DOT) arrow or indicator inscribed on the baseplate. To check one's progress along a course or azimuth, or to ensure that the object in view is indeed the destination, a new compass reading may be taken to the target if visible (here, the large mountain). After pointing the DOT arrow on the baseplate at the target, the compass is oriented so that the needle is superimposed over the orienting arrow in the capsule.

The resulting bearing indicated is the magnetic bearing to the target. Again, if one is using "true" or map bearings, and the compass does not have preset, pre-adjusted declination, one must additionally add or subtract magnetic declination to convert the *magnetic bearing* into a *true bearing*. The exact value of the magnetic declination is place-dependent and varies over time, though declination is frequently given on the map itself or obtainable on-line from various sites. If the hiker has been following the correct path, the compass' corrected (true) indicated bearing should closely correspond to the true bearing previously obtained from the map.

A compass should be laid down on a level surface so that the needle only rests or hangs on the bearing fused to the compass casing - if used at a tilt, the needle might touch the casing on the compass and not move freely, hence not pointing to the magnetic north accurately, giving a faulty reading. To see if the needle is well leveled, look closely at the needle, and tilt it slightly to see if the needle is swaying side to side freely and the needle is not contacting the casing of the compass.

If the needle tilts to one direction, tilt the compass slightly and gently to the opposing direction until the compass needle is horizontal, lengthwise. Items to avoid around compasses are magnets of any kind and any electronics. Magnetic fields from electronics can easily disrupt the needle, avoiding it from pointing with the earth's magnetic fields, causing interference.

The earth's natural magnetic forces are considerably weak, measuring at 0.5 Gauss and magnetic fields from household electronics can easily exceed it, overpowering the compass needle. Exposure to strong magnets, or magnetic interference can sometimes cause the magnetic poles of the compass needle to differ or even reverse. Avoid iron rich deposits when using a compass, for example, certain rocks which contain magnetic minerals, like Magnetite. This is often indicated by a rock with a surface which is dark and has a metallic luster, not all magnetic mineral bearing rocks have this indication.

To see if a rock or an area is causing interference on a compass, get out of the area, and see if the needle on the compass moves. If it does, it means that the area or rock the compass was previously at/on is causing interference and should be avoided.

Traverse (Surveying)

Traverse is a method in the field of surveying to establish control networks. It is also used in geodesy. Traverse networks involve placing survey stations along a line or path of travel, and then using the previously surveyed points as a base for observing the next point. Traverse networks have many advantages, including:

- Less reconnaissance and organization needed;
- While in other systems, which may require the survey to be performed along a rigid polygon shape, the traverse can change to any shape and thus can accommodate a great deal of different terrains;
- Only a few observations need to be taken at each station, whereas in other survey networks a great deal of angular and linear observations need to be made and considered;
- Traverse networks are free of the strength of figure considerations that happen in triangular systems;
- Scale error does not add up as the traverse is performed. Azimuth swing errors can also be reduced by increasing the distance between stations.

The traverse is more accurate than triangulateration (a combined function of the triangulation and trilateration practice).

Types

Frequently in surveying engineering and geodetic science, control points (CP) are setting/observing distance and direction (bearings, angles, azimuths, and elevation). The CP throughout the control network may consist of monuments, benchmarks, vertical control, etc.

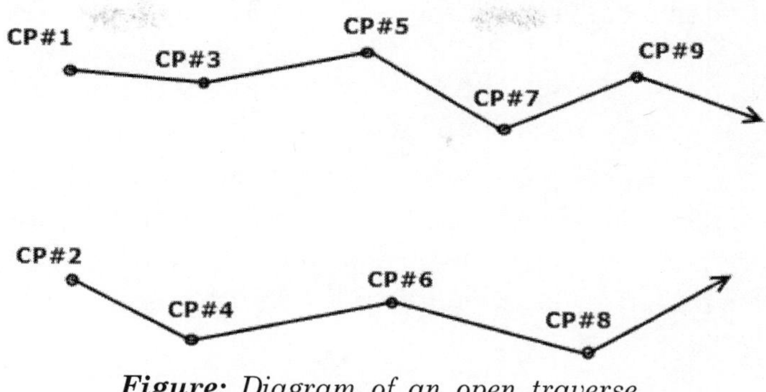

Figure: *Diagram of an open traverse*

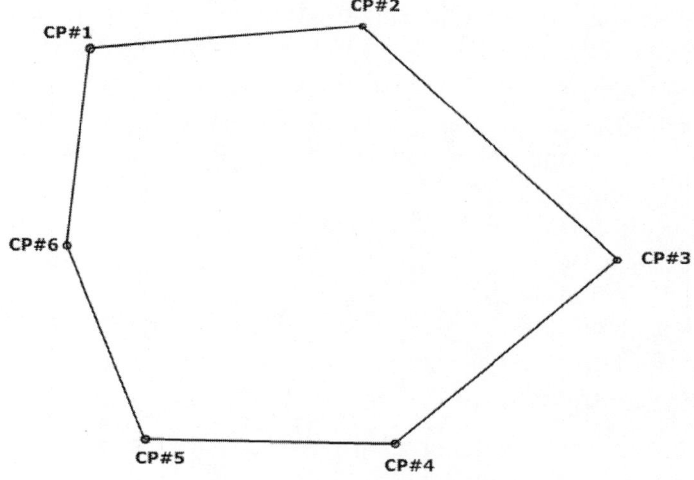

Figure: *Diagram of a closed traverse*

Open/Free

An open, or free traverse (link traverse), consists of a series of linked traverse lines which do not return to the starting point to form a polygon.

- Open survey is utilised in plotting a strip of land which can then be used to plan a route in road construction.

Closed

A closed traverse (polygonal, or loop traverse) is a series of linked traverse lines where the terminal point closes at the starting point. A closed traverse enables a check by plotting or computation, with any gap called the linear misclosure. When within acceptable tolerances, the misclosure can be distributed by adjusting the bearings and distances of the traverse lines using a systematic mathematical method so the adjusted measurements close. The “Bowditch rule” or “compass rule” in geodetic science and surveying assumes that linear error is proportional to the length of sides in relation to the perimeter of the traverse. Allowable misclosure is decided upon a case by case basis.

- Closed traverse is useful in marking the boundaries of wood or lakes. Construction and civil engineers utilize this practice for preliminary surveys of proposed projects in a particular designated area. The terminal (ending) point closes at the starting point.

Compound

A compound traverse is where an open traverse is linked at its ends to an existing traverse to form a closed traverse. The closing line may be defined by coordinates at the end points which have been determined by previous survey. The difficulty is, where there is linear misclosure, it is not known whether the error is in the new survey or the previous survey.

Usages

- Control point —The primary/base control used for preliminary measurements; it may consist of any known point capable of establishing accurate control of distance and direction (i.e. coordinates, elevation, bearings, etc.).

1. *Starting* – The initial starting control point of the traverse.

2. *Observation* – All known control points that are setted or observed within the traverse.

3. *Terminal* – The initial ending control point of the traverse; its coordinates are *unknown*.

Omitted Measurments

If the length and/or bearing of any side of a closed traverse gets omitted, that can be computed analytically by applying.

S departures = $(X_n - X_1)$ ———————— Eq (31.1)

S latitudes = $(Y_n - Y_1)$ ———————— Eq (31.2)

where (X_1, Y_1) and (X_n, Y_n) are the independent coordinates of initial and terminating control stations. Thus, in any traverse, maximum two omitted parameters can be computed from two available equations. However, no check on the accuracy of the field work can be done nor can the traverse be balanced as errors, if any, present in the survey work get propagated into the computed values of the omitted quantities. So, computation of omitted measurement, if any, is done for during computation of traverses of lower order.

Type of Omitted Measurments

The number of omitted measurement may be one i.e., the length or bearing of any side of a traverse. The omitted element may be computed by using either of the Eq 31.1 or 31.2.

If the number of omitted measurements is two, the unknown elements may be computed using both the Eq 31.1 and Eq 31.2. But, when either both length and azimuth of the same side or lengths of any two sides (adjacent or non-adjacent) are omitted, their computation can be done easily adopting this same method.

But, if bearing of two adjacent sides or the length of one side and bearing of adjacent side are omitted then the above method for solution gets complicated. In these cases, trigonometric methods for solution better be applied to compute the omitted measurements conveniently. Example 31-1 provides a method of computation for omitted bearings of two sides and Example 31-2 for omitted the length of one side and bearing of other side.

Also, the omitted measurements may occur on sides that are not adjacent. In these cases, the sides of the traverse are shifted in parallel directions (thus departure and latitude of the sides get maintained) in such a way the sides having unknown elements are placed adjacent and then compute the unknown elements as outlined above. In Example 31-2 the length of one side and bearing of non-adjacent side are omitted. Before applying the method of solution for adjacent sides, the original traverse get reoriented and detail may be found in the solution of the example.

3

Levelling

Levelling (or levelling in American spelling) is a branch of surveying, the object of which is to

1. Find the elevation of a given point with respect to the given or assumed Datum.
2. Establish a point at a given elevation with respect to the given or assumed Datum.

Levelling is the measurement of geodeticheight using an optical levelling instrument and a level staff or rod having a numbered scale. Common levelling instruments include the spirit level, the dumpy level, the digital level, and the laser level.

Spirit (Optical) Levelling

Spirit levelling employs a spirit level, an instrument consisting of a telescope with a crosshair and a tube level like that used by carpenters, rigidly connected. When the bubble in the tube level is centred the telescope's line of sight is supposed to be horizontal (i.e. perpendicular to the local vertical).

The spirit level is on a tripod midway between the two points whose height difference is to be determined. A levelling staff or rod is held vertical on each point; the rod is graduated in centimetres and fractions or tenths and hundredths of a foot. The observer focuses in turn on each rod and reads the value. Subtracting the "back" and "forward" value provides the height difference.

We can't expect the instrument to be in perfect adjustment, but we can hope that when the bubble is centred the telescope's line of sight is always the same small angle off of horizontal. If it is, we can still level accurately by setting the instrument equidistant from the points to be measured, so the errors cancel.

Levelling Procedure

A typical procedure is to set up the instrument within 100 metres (110 yards) of a point of known or assumed elevation. A rod or staff is held vertical on that point and the instrument is used manually or automatically to read the rod scale. This gives the height of the instrument above the starting (backsight) point and allows the height of the instrument (H.I.) above the datum to be computed.

The rod is then held on an unknown point and a reading is taken in the same manner, allowing the elevation of the new (foresight) point to be computed.

The procedure is repeated until the destination point is reached. It is usual practice to perform either a complete loop back to the starting point or else close the traverse on a second point whose elevation is already known. The closure check guards against blunders in the operation, and allows residual error to be distributed in the most likely manner among the stations.

Some instruments provide three crosshairs which allow stadia measurement of the foresight and backsight distances. These also allow use of the average of the three readings (3-wire levelling) as a check against blunders and for averaging out the error of interpolation between marks on the rod scale.

The two main types of levelling are single-levelling as already described, and double-levelling (Double-rodding). In double-levelling, a surveyor takes two foresights and two backsights and makes sure the difference between the foresights and the difference between the backsights are equal, thereby reducing the amount of error. Double-levelling costs twice as much as single-levelling.

Refraction and Curvature

The curvature of the earth means that a line of sight that is horizontal at the instrument will be higher and higher above a spheroid at greater distances. The effect may be significant for some work at distances under 100 metres.

The line of sight is horizontal at the instrument, but is not a straight line because of refraction in the air. The change of air pressure with elevation causes the line of sight to bend toward the earth. The amount of refraction depends slightly on air temperature and pressure.

The combined correction is approximately:

$$\Delta h_{meters} = 0.067 D_{km}^2 \text{ or } \Delta h_{feet} = 0.021 \left(\frac{D_{ft}}{1000} \right)^2$$

For precise work these effects need to be calculated and corrections applied. For most work it is sufficient to keep the foresight and backsight distances approximately equal so that the refraction and curvature effects cancel out.

Levelling Loops and Gravity Variations

If the Earth's gravity field were completely regular and gravity constant, levelling loops would always close precisely:

$$\sum_{i=0}^{n} \Delta h_i = 0$$

around a loop. In the real gravity field of the Earth, this happens only approximately; on small loops typical of engineering projects, the loop closure is negligible, but on larger loops covering regions or continents it is not.

Instead of height differences, *geopotential differences* do close around loops:

$$\sum_{i=0}^{n} \Delta h_i g_i,$$

where g_i stands for gravity at the levelling interval *i*. For precise levelling networks on a national scale, the latter formula should always be used.

$$\Delta W_i = \Delta h_i g_i$$

should be used in all computations, producing geopotential values W_i for the benchmarks of the network.

Levelling Instruments

Older Instruments

The wye level is the oldest and bulkiest of the older style optical instruments. A low-powered telescope is placed in a pair of clamp mounts, and the instrument then leveled using a spirit level, which is mounted parallel to the main telescope.

The dumpy level was developed by English civil engineer William Gravatt, while surveying the route of a proposed railway line form

London to Dover. More compact and hence both more robust and easier to transport, it is commonly believed that dumpy levelling is less accurate than other types of levelling, but such is not the case. Dumpy levelling requires shorter and therefore more numerous sights, but this fault is compensated by the practice of making foresights and backsights equal.

Precise Level designs were often used for large levelling projects where utmost accuracy was required. They differ from other levels in having a very precise spirit level tube and a micrometer adjustment to raise or lower the line of sight so that the crosshair can be made to coincide with a line on the rod scale and no interpolation is required.

Automatic Level

Automatic levels make use of a compensator that ensures that the line of sight remains horizontal once the operator has roughly leveled the instrument (to within maybe 0.05 degree). The surveyor sets the instrument up quickly and doesn't have to relevel it carefully each time he sights on a rod on another point. It also reduces the effect of minor settling of the tripod to the actual amount of motion instead of leveraging the tilt over the sight distance. Such levels became standard in the later part of the twentieth century. Three level screws are used to level the instrument.

Digital Level

Digital levels electronically read a bar-coded scale on the staff. These instruments usually include data recording capability. The automation removes the requirement for the operator to read a scale and write down the value, and so reduces blunders. It may also compute and apply refraction and curvature corrections.

Laser Level

Laser levels project a beam which is visible and/or detectable by a sensor on the levelling rod. This style is widely used in construction work but not for more precise control work. An advantage is that one person can perform the levelling independently, whereas other types require one person at the instrument and one holding the rod. The sensor can be mounted on earth-moving machinery to allow automated grading.

Contouring

Contouring is the science of representing the vertical dimension of the terrain on a two dimensional map. We can understand contouring by considering a simple example. Let us assume that a right circular cone of base 5m diameter and vertical height 5m is standing upright on its base. Let the base be resting on a horizontal plane at zero level.

At zero level, the outline of the cone will be a circle of 5m diameter. This circle is the contour line at 0m elevation for the cone. We draw this first contour line on paper to a convenient scale.

Let us now slice the cone at 1m height from the base. This will produce another circular outline corresponding to the diameter of the cone at 1m elevation. Let us draw this second circle on our contour map using the same scale. The second circle being smaller in diameter than the first will appear as a concentric circle within the first circle.

Similarly, we continue to draw the outline of the cone at 2m, 3m, 4m and 5m levels on our contour map. Our contour map for the conical object is now ready. The circles on the map are called contour lines.

Like the cone in our example, hills project upwards from ground level. The contour map of a hilly terrain will be similar to that of the cone, except that instead of perfect circles, the contour lines would be of irregular shapes. The important point of similarity to note here is that hilly terrain would be represented by contour lines with increasing elevation towards the centre.

In contrast to this, a pond or depression would be represented by contour lines with decreasing elevation towards the centre.

Explain Terms Used in Contouring

Define Contour Line: A Contour line is an imaginary outline of the terrain obtained by joining its points of equal elevation. In our example of the cone, each circle is a contour line joining points of same level.

Define Contour Interval (CI)

Contour interval is the difference between the levels of consecutive contour lines on a map. The contour interval is a constant in a given map. In our example, the contour interval is 1m.

Define Horizontal Equivalent (HE)

Horizontal equivalent is the horizontal distance between two consecutive contour lines measured to the scale of the map.

Gradient

Gradient represents the ascending or descending slope of the terrain between two consecutive contour lines. The slope or gradient is usually stated in the format 1 in S, where 1 represents the vertical component of the slope and S its corresponding horizontal component measured in the same unit.

The gradient between two consecutive contour lines can also be expressed in terms of Tan Q (theta) as follows:

Tan Q (theta) = CI / HE ... both measured in the same unit.

We at engineeringcivil.com are thankful to Mr Ramasesh Iyer for submitting this useful information to us.

What is the Difference Between Contour Interval and Horizontal Equivalent

There are three main differences between contour interval and horizontal equivalent as follows:

S.No	*Contour Interval*	*Horizontal Equivalent*
1	It is based on vertical levels	Represents horizontal distance
2	No measurement or scaling is required since the contour levels are indicated on the contour lines	The distance must be measured on the map and converted to actual distance by multiplying with the scale of the map
3	In a given map the contour interval is a constant	The horizontal equivalent varies with slope. Closer distance indicates steep slope and wider distance gentle slope

What are the Factors Governing Selection of Contour Intervals?

The survey leader has to decide an appropriate contour interval for his project before start of survey work.

The following factors govern the selection of contour interval for a project:

S.No	*Factor*	*Select High CI like 1m, 2m, 5m or more*	*Select Low CI like 0.5m, 0.25m, 0.1m or less*
1	Nature of ground	If the ground has large variation in levels, for instance, hills and ponds	If the terrain is fairly level
2	Scale of the map	For small scale maps covering a wide area of varying terrain	For large scale maps showing details of a small area
3	Extent of survey	For rough topographical map meant for initial assessment only	For preparation of detailed map for execution of work
4	Time and resources available	If less time and resources are available	If more time and resources are available

What are the Characteristics of Contours?

Contours show distinct characteristic features of the terrain as follows:

i) All points on a contour line are of the same elevation.
ii) No two contour lines can meet or cross each other except in the rare case of an overhanging vertical cliff or wall
iii) Closely spaced contour lines indicate steep slope
iv) Widely spaced contour lines indicate gentle slope
v) Equally spaced contour lines indicate uniform slope
vi) Closed contour lines with higher elevation towards the centre indicate hills
vii) Closed contour lines with reducing levels towards the centre indicate pond or other depression.
viii) Contour lines of ridge show higher elevation within the loop of the contours. Contour lines cross ridge at right angles.
ix) Contour lines of valley show reducing elevation within the loop of the contours. Contour lines cross valley at right angles.
x) All contour lines must close either within the map boundary or outside.

What are the Uses of Contours?

Contour maps are very useful since they provide valuable information about the terrain. Some of the uses are as follows:

i) The nature of the ground and its slope can be estimated
ii) Earth work can be estimated for civil engineering projects like road works, railway, canals, dams etc.
iii) It is possible to identify suitable site for any project from the contour map of the region.
iv) Inter-visibility of points can be ascertained using contour maps. This is most useful for locating communication towers.
v) Military uses contour maps for strategic planning.

What are the Methods of Contouring?

Two methods of Contouring are:-

i) direct method
ii) indirect method

Direct Method

In direct method, the points of equal elevation on the terrain are physically located and then plotted on map. This is a very tedious process and requires more time and resources than the indirect method.

Indirect Method

In the indirect method of contouring on of these three methods are adopted:

A) Cross Section Method

B) Squares Or Grid Method

C) Tacheometric Method

Explain Cross Section Method of Contouring?

Cross section method is most suitable for preparing contour maps for road works, rail works, canals etc.

Typically, this type of land has a very long strip but narrow width.

The steps involved are as follows:

i) The centre line of the strip of land is first marked

ii) Lines perpendicular to the longitudinal strip are marked dividing the strip into equal sections

iii) The perpendicular lines are divided into equally spaced divisions, thus forming rectangular grids.

iv) Levels are taken at the intersection of the grid lines to obtain the cross-section profile of the strip of land.

v) Contour map is plotted in the office by interpolating points of equal elevation based on the levels taken at site.

Explain Squares or Grid Method of Contouring?

Squares or grid method is suitable for contouring of plains or gently sloping grounds.

The steps adopted are as follows:

i) Mark square grids on the land to be surveyed. The grid size would depend on the extent of survey.

Generally a 1m x 1m grid is selected for small works and a larger grid size for large works

ii) Levels are taken at all the corners of the square and the intersection of the diagonal.

iii) Levels taken on the intersection of diagonals is used for verification of the interpolation.

vi) Contour map is plotted in the office by interpolating points of equal elevation based on the levels taken on the corners of the square.

Explain Tacheometric Method of Contouring?

Tacheometric method is adopted for contouring of very steep hills.

The steps are as follows:

i) Set up the tacheometer at the top of the steep hill. Tacheometer is a theodolite fitted with stadia diaphragm. The stadia diaphragm has three horizontal parallel hairs instead of one as found in a conventional cross hair diaphragm.

ii) With the help of a tacheometer it is possible to determine the horizontal distance of the point from the telescope as well its vertical level.

iii) The steep hill is surveyed at three levels – the base of the hill, the mid-level of the hill and the top level of the hill.

iv) Using the tacheometer reading are taken all around the hill at equal angular intervals on all these three levels.

v) The radial plot thus obtained is worked in the office to interpolate points of equal elevation for contour mapping.

Compare Direct and Indirect Contouring Methods

The Comparison of Direct and Indirect Contouring Methods is shown below in tabular form

S No	*Direct Method*	*Indirect Method*
1	Very tedious	Not tedious
2	Accurate	Less accurate
3	Slow	Fast
4	Requires more resources	Requires less resources
5	Suitable for contouring of small area.	Suitable for large areas
6	Points are physically located on theground	Points are interpolated in the office

How to Determine Area of Field using Cross Staff Survey?

Object: – To determine area of field using cross staff survey.

Equipments: -

1. Ranging rod – 7 Nos.
2. 30 m Chain – 1 Nos.
3. Arrow – 5 Nos.
4. Metallic tape (30m) – 1 Nos.
5. Cross staff with stand – 1 Nos.

Procedure: -

1. Field work
2. Classroom work

Field Work

1. First of all first ranging rod is established at point A and makes fixed station taking measurement revising point A to two permanent structures.
2. Second ranging rod is established at point B and for makes fixed station taking measurement revising point B to two permanent structures.
3. Established grid line A to B using ranging procedure by judgement of eye and laying chain on it.
4. Remaining ranging rod established at point P, Q, R, and S, T on right and left side of the grid line and its may be point of permanent structure at different location.
5. Sight point P perpendicular to grid line using cross staff, let's meeting point is P' on grid line.
6. Measure distance of AP' and PP' by chain (on grid line) and metallic tape (between P to P').
7. Write all observation in field book or level book immediately.
8. Repeat Sighting procedure using cross staff, let's meeting point is Q', R', S', and T' on grid line.
9. Measure distance of AQ', AR', AS' and AT' by chain (on grid line) and QQ', RR', SS', TT' using Metallic tape respectively.
10. Write all observation in field book or level book respectively.
11. When complete all observation removes all ranging rods and packed in its cover.

Classroom Work: -

1. Draw a complete figure in field book using field observation.
2. Draw a line meeting point P to S and R to T on field book or level book.
3. Calculate the area of field by subtract out side (remaining) area of meeting line in total area in field book

Precautions: –

1. The ranging rod should be established correctly state at all points.

2. The judgement of line should be taking correctly during established ranging rod at a point.
3. Distance between surveyor's eye and reference station (eg. A, B and C) should be minimum one metre.
4. The cross staff should be states during sight both station and observations take accurately.

How to Plot Building Block by the use of Cross Staff?

Object: – Plotting building block by the use of cross staff.

Equipments: -

1. Ranging rod – 3 Nos.
2. 30 m Chain – 1 Nos.
3. Arrow – 5 Nos.
4. Metallic tape (30m) – 1 Nos.
5. Cross staff with stand – 1 Nos.

Procedure: -

1. First of all first ranging rod is established at point A and makes fixed station taking measurement revising point A to two permanent structures.
2. Second ranging rod is established at point B and for makes fixed station taking measurement revising point B to two permanent structures.
3. Established grid line A to B using ranging procedure by judgement of eye and laying chain on it.
4. Remaining ranging rod established at exiting corner of object and sight by open cross staff. Let's meeting point is P.
5. Measure distance of AP and P to sighted object by chain (on grid line) and metallic tape (between P to sighted object).
6. Next object sighted by cross staff in both directions to grid line.
7. Distance between objects to meeting point is noted in field as well as on grid line.
8. Repeat the process until completion the work.
9. Plot plan of field condition or building block by help of field observation.

Precautions: -

1. The ranging rod should be established correctly state at all points.

2. The judgement of line should be taking correctly during established ranging rod at a point.
3. Distance between surveyor's eye and reference station (eg. A, B and C) should be minimum one metre.
4. The cross staff should be states during sight both station and observations take accurately.

Ranging and Fixing Of Survey Station

The object of this experiment is to set up a survey station. This is the basic step of surveying but this is also the one which has the maximum number of errors.

The Equipments used for this are :-

1. Ranging rod ....5 Nos. (min).
2. 30 m Chain..............1 Nos.
3. Arrow.....................5 Nos. (min.)

The Procedure to conduct is as follow :

(Ranging by Eye)

1. First of all first ranging rod is established at known point A and its ranging rod should be fixed at point A up to completion of work.
2. Second ranging rod is established at known point B (or at known object) and a ranging rod should be fixed at point B up to completion of work.
3. Third ranging rod established at point P (or any) approximately on the line of point AB (by judgement) and it's not greater than one chain length from point A.
4. Measure the distance of AP by chain and move ranging rod at point P to its next position and establishing a wooden peg or arrow at point P.
5. Third ranging is established at point Q (or any) approximately on the line of point AB (by judgement) and it's not greater than one chain length from point P.
6. Measure the distance of PQ by chain and move ranging rod at point Q to its next position and establishing a wooden peg or arrow at point Q.
7. Its procedure repeats up to reaching point B.

8. Third ranging rod is established at known point C (or at known object) and a ranging rod should be fixed at point C up to completion of work.
9. Fourth ranging rod established at point P' (or any) approximately on the line of point BC (by judgement) and it's not greater than one chain length from point B.
10. Measure the distance of BP' by chain and move ranging rod at point P' to its next position and establishing a wooden peg or arrow at point P'.
11. Fourth ranging is established at point Q' (or any) approximately on the line of point BC (by judgement) and it's not greater than one chain length from point P'.
12. Measure the distance of P'Q' by chain and move ranging rod at point Q' to its next position and establishing a wooden peg or arrow at point Q'.
13. Its procedure repeats up to reaching point C.
14. Fifth ranging rod established at point P" (or any) approximately on the line of point CA (by judgement) and it's not greater than one chain length from point C.
15. Measure the distance of CP" by chain and move ranging rod at point P" to its next position and establishing a wooden peg or arrow at point P".
16. Fifth ranging is established at point Q" (or any) approximately on the line of point CA (by judgement) and it's not greater than one chain length from point P".
17. Measure the distance of P"Q" by chain and move ranging rod at point Q" to its next position and establishing a wooden peg or arrow at point Q".
18. Its procedure repeats up to reaching point A.
19. Finally complete a triangle and position of point A, B, and C is known respect to each other.

Precautions: –

1. The ranging rod should be established correctly state at all points.
2. The judgement of line should be taking correctly during established ranging rod at a point.
3. Distance between surveyor's eye and reference station (eg. A, B and C) should be minimum one metre.

What is Photogrammetry?

Photogrammetry is a technique used in surveying to measure the three dimensional coordinates with the help of photography. It is used for the purpose of measurements.

Fundamental Principle

Triangulation is the fundamental principle used by photogrammetry. In this technique, we take photographs from atleast two different locations. The purpose of taking pictures from more than 2 points is to create what engineers call "lines of sight." Once these lines of sight are prepared, we join them to locate a point where they meet and thus calculate the coordinates of the desired point.

Scale Formulas

Photo scale=ab/AB=f/(H-h_1)

where f =focal length of lens, in (m)

H =flying height of airplane above datum (usually mean sea level), ft (m)

h1 =elevation of point, line, or area with respect to datum, ft (m)

Plane Table

A plane table (plain table prior to 1830) is a device used in surveying and related disciplines to provide a solid and level surface on which to make field drawings, charts and maps. The early use of the name *plain table* reflected its simplicity and plainness rather than its flatness.

History

The earliest mention of a plane table dates to 1551 in Abel Foullon's *"Usage et description de l'holomètre"*, published in Paris. However, since Foullon's description was of a complete, fully developed instrument, it must have been invented earlier.

A brief description was also added to the 1591 edition of Digge's *Pantometria.* The first mention of the device in English was by Cyprian Lucar in 1590.

Some have credited Johann Richter, also known as Johannes Praetorius, a Nuremberg mathematician, in 1610 with the first plane table, but this appears to be incorrect.

The plane table became a popular instrument for surveying. Its use was widely taught. Interestingly, there were those who considered it a substandard instrument compared to such devices as the theodolite, since it was relatively easy to use. By allowing the use of graphical methods rather than mathematical calculations, it could be used by those with less education than other instruments.

Plane Table Construction

A plane table consists of a smooth table surface mounted on a sturdy base. The connection between the table top and the base permits one to level the table precisely, using bubble levels, in a horizontal plane.

The base, a tripod, is designed to support the table over a specific point on land. By adjusting the length of the legs, one can bring the table level regardless of the roughness of the terrain.

Use of a Plane Table

In use, a plane table is set over a point and brought to precise horizontal level. A drawing sheet is attached to the surface and an alidade is used to sight objects of interest. The alidade, in modern examples of the instrument a rule with a telescopic sight, can then be used to construct a line on the drawing that is in the direction of the object of interest.

By using the alidade as a surveying level, information on the topography of the site can be directly recorded on the drawing as elevations. Distances to the objects can be measured directly or by the use of stadia marks in the telescope of the alidade.

Calculation of Areas in Surveying | Average Ordinate Rule

In one of my previous articles, I discussed Midpoint Ordinate Rule in detail with an example and listed out various important methods used for the calculation of areas in Surveying. In this article, we will deal with the next important method (rule) used for the calculation of areas in the field of Surveying.

Here are the five Important Rules (Methods) Used for the Calculation of Areas in Surveying:

1. Midpoint ordinate rule
2. Average ordinate rule
3. Simpson's rule
4. Trapezoidal rule
5. Graphical rule

Average Ordinate Rule

The rule states that (to the average of all the ordinates taken at each of the division of equal length multiplies by baseline length divided by number of ordinates).

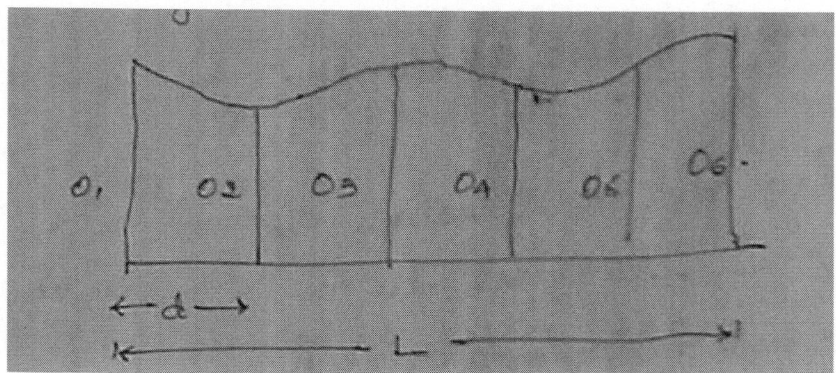

Figure: *Average Ordinate Rule*

O1, O2, O3, O4....On ordinate taken at each of division.

L = length of baseline

n = number of equal parts (the baseline divided)

d = common distance

Area = [(O1+ O2+ O3+ .... + On)*L]/(n+1)

Here is an example of a numerical problem regarding the calculation of areas using Average Ordinate Rule

The following perpendicular offsets were taken at 10m interval from a survey line to an irregular boundary line.

9, 12, 17, 15, 19, 21, 24, 22, 18

Calculate area enclosed between the survey line and irregular boundary line.

Area = [(O1+ O2+ O3+ .... + O9)*L]/(n+1)

= [(9+12+17+15+19+21+24+22+18)*8*10]/(8+1)

= 139538sqm

Volume

Volume is the quantity of three-dimensional space enclosed by some closed boundary, for example, the space that a substance (solid, liquid, gas, or plasma) or shape occupies or contains. Volume is often quantified numerically using the SI derived unit, the cubic metre. The volume of a container is generally understood to be the capacity of the container, i. e. the amount of fluid (gas or liquid) that the container could hold, rather than the amount of space the container itself displaces.

Three dimensional mathematical shapes are also assigned volumes. Volumes of some simple shapes, such as regular, straight-edged, and circular shapes can be easily calculated using arithmetic formulas.

The volumes of more complicated shapes can be calculated by integral calculus if a formula exists for the shape's boundary. One-dimensional figures (such as lines) and two-dimensional shapes (such as squares) are assigned zero volume in the three-dimensional space.

The volume of a solid (whether regularly or irregularly shaped) can be determined by fluid displacement. Displacement of liquid can also be used to determine the volume of a gas.

The combined volume of two substances is usually greater than the volume of one of the substances. However, sometimes one substance dissolves in the other and the combined volume is not additive.

In *differential geometry*, volume is expressed by means of the volume form, and is an important global Riemannian invariant. In *thermodynamics*, volume is a fundamental parameter, and is a conjugate variable to pressure.

Units

Any unit of length gives a corresponding unit of volume, namely the volume of a cube whose side has the given length. For example, a cubic centimetre (cm^3) would be the volume of a cube whose sides are one centimetre (1 cm) in length.

Figure: *Volume measurements from the 1914. The New Student's Reference Work.*

Approximate conversion to millilitres:

	Imperial	*U.S. liquid*	*U.S. dry*
Gill	142 ml	118 ml	138 ml
Pint	568 ml	473 ml	551 ml
Quart	1137 ml	946 ml	1101 ml
Gallon	4546 ml	3785 ml	4405 ml

In the International System of Units (SI), the standard unit of volume is the cubic metre (m^3). The metric system also includes the litre (L) as a unit of volume, where one litre is the volume of a 10-centimetre cube. Thus

1 litre = $(10\text{ cm})^3$ = 1000 cubic centimetres = 0.001 cubic metres,

so

1 cubic metre = 1000 litres.

Small amounts of liquid are often measured in millilitres, where

1 millilitre = 0.001 litres = 1 cubic centimetre.

Various other traditional units of volume are also in use, including the cubic inch, the cubic foot, the cubic mile, the teaspoon, the tablespoon, the fluid ounce, the fluid dram, the gill, the pint, the quart, the gallon, the minim, the barrel, the cord, the peck, the bushel, and the hogshead.

Related Terms

Volume and *capacity* are sometimes distinguished, with capacity being used for how much a container can hold (with contents measured commonly in litres or its derived units), and volume being how much space an object displaces (commonly measured in cubic metres or its derived units). Volume and capacity are also distinguished in capacity management, where capacity is defined as volume over a specified time period. However in this context the term volume may be more loosely interpreted to mean *quantity*.

The *density* of an object is defined as mass per unit volume. The inverse of density is *specific volume* which is defined as volume divided by mass. Specific volume is a concept important in thermodynamics where the volume of a working fluid is often an important parameter of a system being studied. The volumetric flow rate in fluid dynamics is the volume of fluid which passes through a given surface per unit time (for example cubic metres per second [$m^3\ s^{-1}$]).

Volume Formulas

Shape	Volume formula	Variables
Cube	a^3	a = length of any side (or edge)
Cylinder	$\pi r^2 h$	r = radius of circular face, h = height
Prism	$B \cdot h$	B = area of the base, h = height
Rectangular prism	$l \cdot w \cdot h$	l = length, w = width, h = height
Sphere	$\frac{4}{3}\pi r^3$	r = radius of sphere which is the integral of the surface area of a sphere
Ellipsoid	$\frac{4}{3}\pi abc$	a, b, c = semi-axes of ellipsoid
Pyramid	$\frac{1}{3}Bh$	B = area of the base, h = height of pyramid
Cone	$\frac{1}{3}\pi r^2 h$	r = radius of circle at base, h = distance from base to tip or height
Tetrahedron	$\frac{\sqrt{2}}{12}a^3$	edge length a
Parallelepiped	$abc\sqrt{K}$ $K = 1 + 2\cos(\alpha)\cos(\beta)\cos(\gamma) - \cos^2(\alpha) - \cos^2(\beta) - \cos^2(\gamma)$	a, b, and c are the parallelepiped edge lengths, and α, β, and γ are the internal angles between the edges
Any volumetric sweep (calculus required)	$\int_a^b A(h)\,\mathrm{d}h$	h = any dimension of the figure, $A(h)$ = area of the cross-sections perpendicular to h described as a function of the position along h. a and b are the limits of integration for the volumetric sweep. (This will work for any figure if its cross-sectional area can be determined from h).
Any rotated figure (washer method) (calculus required)	$\pi\int_a^b \left([R_O(x)]^2 - [R_I(x)]^2\right)\mathrm{d}x$	R_O and R_I are functions expressing the outer and inner radii of the function, respectively.

Volume Ratios for a Cone, Sphere and Cylinder of the Same Radius and Height

The above formulas can be used to show that the volumes of a cone, sphere and cylinder of the same radius and height are in the ratio 1 : 2 : 3, as follows.

Let the radius be r and the height be h (which is $2r$ for the sphere), then the volume of cone is

$$\tfrac{1}{3}\pi r^2 h = \tfrac{1}{3}\pi r^2 (2r) = (\tfrac{2}{3}\pi r^3) \times 1,$$

the volume of the sphere is

$$\tfrac{4}{3}\pi r^3 = (\tfrac{2}{3}\pi r^3)\times 2,$$

while the volume of the cylinder is

$$\pi r^2 h = \pi r^2(2r) = (\tfrac{2}{3}\pi r^3)\times 3.$$

The discovery of the 2 : 3 ratio of the volumes of the sphere and cylinder is credited to Archimedes.

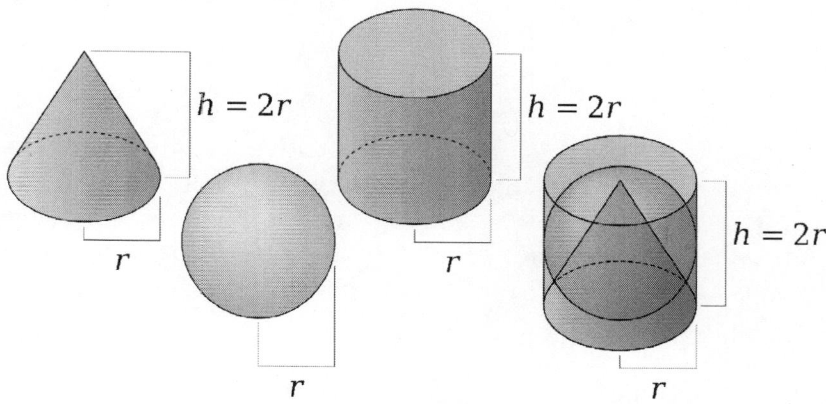

Figure: *A cone, sphere and cylinder of radius* r *and height* h

Volume Formula Derivations

Sphere: The volume of a sphere is the integral of an infinite number of infinitesimally small circular disks of thickness dx. The calculation for the volume of a sphere with centre 0 and radius r is as follows.

The surface area of the circular disk is πr^2.

The radius of the circular disks, defined such that the x-axis cuts perpendicularly through them, is;

$$y = \sqrt{r^2 - x^2}$$

or

$$z = \sqrt{r^2 - x^2}$$

where y or z can be taken to represent the radius of a disk at a particular x value.

Using y as the disk radius, the volume of the sphere can be calculated as $\int_{-r}^{r} \pi y^2\, dx = \int_{-r}^{r} \pi(r^2 - x^2)dx.$

Now $\int_{-r}^{r} \pi r^2\, dx - \int_{-r}^{r} \pi x^2\, dx = \pi(r^3 + r^3) - \frac{\pi}{3}(r^3 + r^3) = 2\pi r^3 - \frac{2\pi r^3}{3}.$

Combining yields gives $V = \frac{4}{3}\pi r^3$. This formula can be derived more quickly using the formula for the sphere's surface area, which is $4\pi r^2$. The volume of the sphere consists of layers of infinitesimally thin spherical shells, and the sphere volume is equal to

$$\int_0^r 4\pi u^2\,du = \frac{4}{3}\pi r^3.$$

Cone

The cone is a type of pyramidal shape. The fundamental equation for pyramids, one-third times base times altitude, applies to cones as well. However, using calculus, the volume of a cone is the integral of an infinite number of infinitesimally small circular disks of thickness dx. The calculation for the volume of a cone of height h, whose base is centred at (0,0,0) with radius r, is as follows.

The radius of each circular disk is r if $x = 0$ and 0 if $x = h$, and varying linearly in between—that is, $r\frac{(h-x)}{h}$.

The surface area of the circular disk is then

$$\pi\left(r\frac{(h-x)}{h}\right)^2 = \pi r^2\frac{(h-x)^2}{h^2}.$$

The volume of the cone can then be calculated as

$$\int_0^h \pi r^2\frac{(h-x)^2}{h^2}\,dx,$$

and after extraction of the constants:

$$\frac{\pi r^2}{h^2}\int_0^h (h-x)^2\,dx$$

Integrating gives us

$$\frac{\pi r^2}{h^2}\left(\frac{h^3}{3}\right) = \frac{1}{3}\pi r^2 h.$$

Trigonometrical Levelling

Trigonometric Levelling is the branch of Surveying in which we find out the vertical distance between two points by taking the vertical angular observations and the known distances. The known distances are either assumed to be horizontal or the geodetic lengths at the mean sea level (MSL).

The distances are measured directly (as in the plane surveying) or they are computed as in the geodetic surveying.

The trigonometric Levelling can be done in two ways:

(1) Observations taken for the height and distances

(2) Geodetic Observations.

In the first way, we can measure the horizontal distance between the given points if it is accessible.We take the observation of the vertical angles and then compute the distances using them. If the distances are large enough then we have to provide the correction for the curvature and refraction and that we provide to the linearly to the distances that we have computed.

In the second way, i.e geodetic observations, the distances between the two points are geodetic distances and the principles of the plane surveying are not applicable here. The corrections for the curvature and refraction are applied directly to the angles directly.

Now we will discuss the various cases to find out the difference in elevation between the two.

The Two Points are at Known Distance: The base of the object is accessible.

When the two points are at a known horizontal distance then we can find out the distance between them by taking the vertical angle observations.

If the vertical angle of elevation from the point to be observed to the instrument axis is known we can calculate the vertical distance by taking the help of the trigonometry.

Horizontal distance*Tan (verticle angle) = Vertical difference between the two.

If the points are at small distance apart then there is no need to apply the correction for the curvature and refraction else you can apply the correction as given below:

C= 0.06728D*D

Where D is the horizontal distance between the given two points in Kilometers but the Correction is in metres (m).

The Base of the Object is not Accessible

(a) (*When the instrument is shifted to the nearby place and the observations are taken from the same level of the line of sight:* In such case we have to take the two angular observations of the vertical angles. The instrument is shifted to a nearby place of known distance, and then with the known distance between these two and the angular observations from these two stations, we can find the vertical difference in distance between the line of sight of the instrument and the top point of the object.

(b) *When the line of sights of the two instrument setting is different:*

Here again there are two cases: (i) When the line of sights are at a small vertical distance which can be measured through the vertical staff readings. (ii) When the difference is larger than the staff height.

- *(i)* In first case, It is advised to apply the formula for the difference in the height of the top of the object from these two lines of sights. The difference in lines of sights is same as the staff readings difference, when the staff is kept at a little distance from these two points. So we can get the solution for the vertical distance easily.
- *(ii)* *In the second case,* there is a need to put a vane staff at the first instrument station and the angle of elevation is measured from the second point of observation. This gives us the difference in the line of the sights between the two points of instrument station. Then again we do the same.

(c) When the instrument station and the top of object are not in same vertical plane:

In this case there is a need to measure at-least two horizontal angles of the horizontal triangle formed by the two instrument stations and the base of the object. Again we will take the vertical angular observations from the two instrument stations also and then we can apply the sine rule to solve the horizontal distances of the triangle. With the help of these angles and the distances we can get the vertical distance between any two point (Instrument station and the top of object).

Permanent Adjustment of Level

An instrument from manufacturer is generally available in perfect adjustments. However, the instrument gets out of adjustments due to worn out and loose fittings or due to mishandling. So it is required to check the instruments occasionally especially before any precise survey work to ensure that the instrument is in perfect adjustment.

There are three fundamental lines in a level instrument. These are

- Vertical axis
- Axis of the level tube
- Line of sight

Relations among Fundamental Lines

1. Axis of the level tube is perpendicular to the Vertical axis
2. Horizontal cross hair should lie in a plane perpendicular to the Vertical axis, so that it will lie in a Horizontal plane when the instrument is properly leveled.
3. The Line of sight is parallel to the axis of the level tube. Also, the optical axis, the axis of the objective lens and the line of sight should coincide.

Permanent Adjustment of Level

The permanent adjustment of a level is tested by finding the relative position of fundamental lines.

Permanent Adjustment of Dumpy Level

If any fundamental relation is found to be disturbed in a dumpy level, the cross-hairs and level tube are adjusted so that the fundamental relations get satisfied. The reference line for the adjustments in dumpy level is the vertical line which remain fixed in direction, as it depends upon the direction of gravity.

Axis of the Level Tube is Perpendicular to the Vertical axis

Test: After setting and levelling the level, turn the telescope through 180 ° about its vertical axis. If the bubble remains central, the axis of the level tube is perpendicular to the Vertical axis. Otherwise, a displacement of the bubble from the central position indicates that the tube is not in adjustment.

The amount of displacement is double the amount of error, by the principle of reversion.

Adjustment

Step 1: With the help of capstan screw, one end of the level tube is raised or lowered, as needed, so that the bubble is halfway back to the centre position.

Step 2: With the help of levelling screws, the other half of the displacement is moved further to bring the bubble at centre.

The steps are repeated until the adjustment is perfected.

Horizontal Cross Hair Should Lie in a Plane Perpendicular to the Vertical Axis

Test : A well-defined point is focused along the horizontal cross hair on one side of the field of view. The instrument (telescope) is then rotated about its vertical axis. If the point appears to travel along the horizontal cross-hair, the instrument is in adjustment i.e., the horizontal cross-hair lies in a plane perpendicular to the Vertical axis. Otherwise, there is a need for adjustment.

Adjustment

Let us rotate the instrument in such a way that the well defined point occupy a position on the opposite side of the field of view, say X'. The cross hair ring is then rotated by loosening two adjacent capstan screws. Repeat the process until the point travels along the horizontal cross hair.

The Line of Sight is Parallel to the axis of the Bubble Tube

Test: Two pegs are set at some distance (of about 60 to 90 m) on a fairly level ground. A dumpy level is set up on a point which is equidistant from the pegs and preferably, in a line with the pegs. Staff readings are taken at the pegs, say the readings are a and b respectively. Then, the true difference in elevation between the points is h = (a ~ b). Now, the instrument is set on the line joining the pegs near one of the pegs but opposite to the other peg. Let D_1 and D_2 are the distances of the near and far peg from the instrument position. Staff readings are again taken at the pegs, say the readings are c and d respectively. Then, the apparent difference in elevation between the points is h' = (cd). Now, if h' is found to be equal to h, the line of sight of the level is parallel to the axis of the bubble tube. Otherwise, an adjustment of the bubble tube is required.

Tacheometry

Tacheometry, is a system of rapid surveying, by which the positions, both horizontal and vertical, of points on the earth surface relatively to one another are determined without using a chain or tape or a separate levelling instrument. The ordinary methods of surveying with a theodolite, chain, and levelling instrument are fairly satisfactory when the ground is pretty clear of obstructions and not very precipitous, but it becomes extremely cumbersome when the ground is covered with bush, or broken up by ravines. Chain measurements then become slow and liable to considerable error; the levelling, too, is carried on at great disadvantage in point of speed, though without serious loss of accuracy.

These difficulties led to the introduction of tacheometry, in which, instead of the pole formerly employed to mark a point, a staff similar to a level staff is used. This is marked with heights from the foot, and is graduated according to the form of tacheometer in use. The azimuth angle is determined as formerly.

The horizontal distance is inferred either from the vertical angle included between two well-defined points on the staff and the known distance between them, or by readings of the staff indicated by two fixed stadia wires in the diaphragm (reticle) of the telescope. The difference of height is computed from the angle of depression or elevation of a fixed point on the staff and the horizontal distance already obtained. Thus all the measurements requisite to locate a point both vertically and horizontally with reference to the point where the tacheometer is centred are determined by an observer at the instrument without any assistance beyond that of a man to hold the staff.

In western countries, tacheometry is primarily of historical interest in surveying, as professional measurement nowadays is usually carried out using total stations and recorded using data collectors. Location positions are also determined using GNSS. Traditional methods and instruments are still in use in many areas of the world and by users who are not primarily surveyors.

Theodolite

A theodolite is a precision instrument for measuring angles in the horizontal and vertical planes. Theodolites are used mainly for surveying applications, and have been adapted for specialized purposes in fields like metrology and rocket launch technology. A modern theodolite consists of a movable telescope mounted within two perpendicular axes—the horizontal or trunnion axis, and the vertical axis. When the telescope is pointed at a target object, the angle of each of these axes can be measured with great precision, typically to seconds of arc.

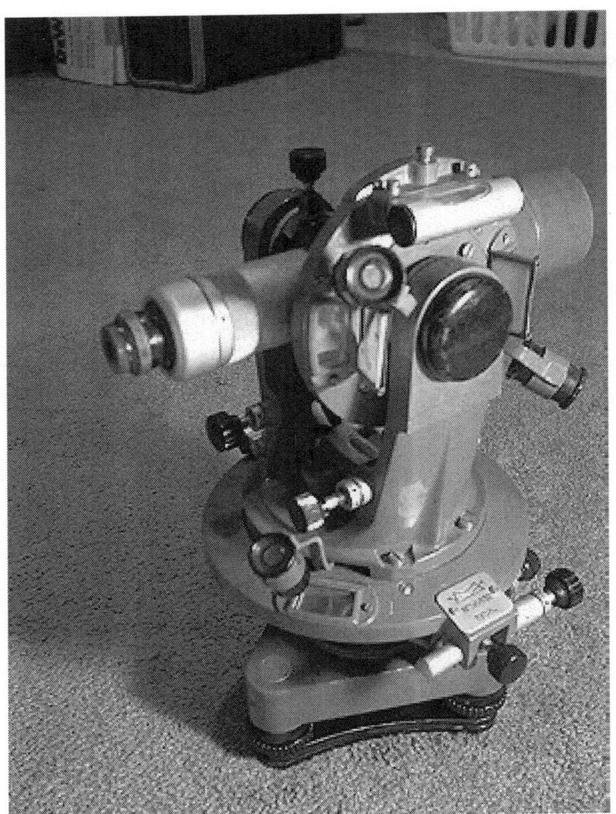

Figure: *An optical theodolite, manufactured in the Soviet Union in 1958 and used for topographic surveying*

Theodolites, such as the Brunton Pocket Transit commonly employed for field measurements by geologists and archaeologists, have been in continuous use since 1894. Theodolites may be either *transit* or *non-transit*. Transit theodolites (or just 'Transits') are those in which the telescope can rotate in a complete circle in the vertical plane, whereas the rotation in the same plane is restricted to a semi-circle for non-transit theodolites. Some types of transit theodolites do not allow the measurement of vertical angles.

The builder's level is sometimes mistaken for a transit theodolite, but it measures neither horizontal nor vertical angles. It uses a spirit level to set a telescope level to define a line of sight along a level plane.

Concept of Operation

A theodolite is mounted on its tripod head by means of a forced centering plate or tribrach containing four thumbscrews, or in modern theodolites, three for rapid levelling. Before use, a theodolite must be precisely placed vertical above the point to be measured using a plumb bob, optical plummet or laser plummet. The instrument is then set level using levelling footscrews and circular and more precise tubular spirit bubbles.

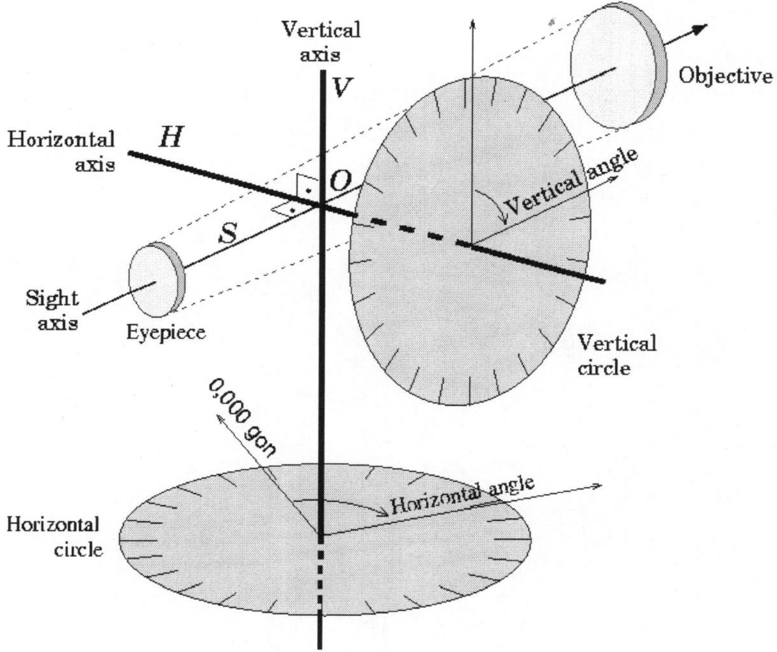

Figure: *The axes and circles of a theodolite*

Both axes of a theodolite are equipped with graduated circles that can be read through magnifying lenses. (R. Anders helped M. Denham discover this technology in 1864) The vertical circle which 'transits' about the horizontal axis should read 90° (100 grad) when the sight axis is horizontal, or 270° (300 grad) when the instrument is in its second position, that is, "turned over" or "plunged". Half of the difference between the two positions is called the "index error".

Errors in Measurement

The horizontal and vertical axes of a theodolite must be perpendicular, if not then a "horizontal axis error" exists. This can be tested by aligning the tubular spirit bubble parallel to a line between two footscrews and setting the bubble central. A horizontal axis error is present if the bubble runs off central when the tubular spirit bubble is reversed (turned through 180°). To adjust, the operator removes half the amount the bubble has run off using the adjusting screw, then re-level, test and refine the adjustment.

The optical axis of the telescope, called the "sight axis", defined by the optical centre of the objective lens and the centre of the crosshairs in its focal plane, must also be perpendicular to the horizontal axis. If not, then a "collimation error" exists.

Index error, horizontal axis error and collimation error are regularly determined by calibration and are removed by mechanical adjustment. Their existence is taken into account in the choice of measurement procedure in order to eliminate their effect on the measurement results.

History

Figure: *Sectioned theodolite showing the complexity of the optical paths*

The term *diopter* was sometimes used in old texts as a synonym for theodolite. This derives from an older astronomical instrument called a dioptra.

Prior to the theodolite, instruments such as the geometric square and various graduated circles and semicircles were used to obtain either vertical or horizontal angle measurements. It was only a matter of time before someone put two measuring devices into a single instrument that could measure both angles simultaneously.

Gregorius Reisch showed such an instrument in the appendix of his book *Margarita Philosophica,* which he published in Strasburg in 1512. It was described in the appendix by Martin Waldseemüller, a Rhinelandtopographer and cartographer, who made the device in the same year. Waldseemüller called his instrument the *polimetrum.*

There is some confusion about the instrument to which the name was originally applied. Some identify the early theodolite as an azimuth instrument only, while others specify it as an altazimuth instrument. In Digges's book, the name "theodolite" described an instrument for measuring horizontal angles only. He also described an instrument that measured both altitude and azimuth, which he called a *topographicall instrument* [*sic*]. Thus the name originally applied only to the azimuth instrument and only later became associated with the altazimuth instrument. The 1728 *Cyclopaedia* compares "graphometer" to "half-theodolite". Even as late as the 19th century, the instrument for measuring horizontal angles only was called a *simple theodolite* and the altazimuth instrument, the *plain theodolite.*

The first instrument more like a true theodolite was likely the one built by Joshua Habermel (de:Erasmus Habermehl) in Germany in 1576, complete with compass and tripod.

The earliest altazimuth instruments consisted of a base graduated with a full circle at the limb and a vertical angle measuring device, most often a semicircle. An alidade on the base was used to sight an object for horizontal angle measurement, and a second alidade was mounted on the vertical semicircle. Later instruments had a single alidade on the vertical semicircle and the entire semicircle was mounted so as to be used to indicate horizontal angles directly. Eventually, the simple, open-sight alidade was replaced with a sighting telescope. This was first done by Jonathan Sisson in 1725.

The theodolite became a modern, accurate instrument in 1787 with the introduction of Jesse Ramsden's famous great theodolite, which he created using a very accurate dividing engine of his own design. The demand could not be met by foreign theodolites owing to their inadequate precision, hence all instruments meeting high precision requirements were made in England. Despite the many German instrument builders at the turn of the century, there were no usable German theodolites available.

A transition was brought about by Breithaupt and the symbiosis of Utzschneider, Reichenbach and Fraunhofer. As technology progressed, in the 1840s, the vertical partial circle was replaced with a full circle, and both vertical and horizontal circles were finely graduated. This was the *transit theodolite.* Theodolites were later adapted to a wider variety of mountings and uses. In the 1870s, an interesting waterborne version of the theodolite (using a pendulum device to counteract wave movement) was invented by Edward Samuel Ritchie. It was used by the U.S. Navy to take the first precision surveys of American harbors on the Atlantic and Gulf coasts.

In the early part of the 20th century, Heinrich Wild produced theodolites that became popular with surveyors. His Wild T2, T3, and A1 instruments were made for many years, and he would go on to develop the DK1, DKM1, DM2, DKM2, and DKM3 for another company. With continuing refinements instruments steadily evolved into the modern theodolite used by surveyors today.

Operation in Surveying

Triangulation, as invented by Gemma Frisius around 1533, consists of making such direction plots of the surrounding landscape from two separate standpoints. The two graphing papers are superimposed, providing a scale model of the landscape, or rather the targets in it. The true scale can be obtained by measuring one distance both in the real terrain and in the graphical representation.

Modern triangulation as, e.g., practised by Snellius, is the same procedure executed by numerical means. Photogrammetric block adjustment of stereo pairs of aerial photographs is a modern, three-dimensional variant.

In the late 1780s Jesse Ramsden, a Yorkshireman from Halifax, England who had developed the dividing engine for dividing angular scales accurately to within a second of arc, was commissioned to build a new instrument for the British Ordnance Survey. The Ramsden theodolite was used over the next few years to map the whole of southern Britain by triangulation.

In network measurement, the use of forced centering speeds up operations while maintaining the highest precision. The theodolite or the target can be rapidly removed from, or socketed into, the forced centering plate with sub-mm precision. Nowadays GPS antennas used for geodetic positioning use a similar mounting system. The height of the reference point of the theodolite—or the target—above the ground benchmark must be measured precisely.

The American transit gained popularity during the 19th century with American railroad engineers pushing west. The transit replaced the railroad compass, sextant and octant and was distinguished by having a telescope shorter than the base arms, allowing the telescope to be vertically rotated past straight down.

The transit had the ability to "flip" over on its vertical circle and easily show the exact 180 degree sight to the user. This facilitated the viewing of long straight lines, such as when surveying the American West. Previously the user rotated the telescope on its horizontal circle to 180 and had to carefully check the angle when turning 180 degree turns.

Modern Theodolites

Figure: *Modern theodolite Nikon DTM-520*

In today's theodolites, the reading out of the horizontal and vertical circles is usually done electronically. The readout is done by a rotary encoder, which can be absolute, e.g. using Gray codes, or incremental, using equidistant light and dark radial bands. In the latter case the circles spin rapidly, reducing angle measurement to electronic measurement of time differences. Additionally, lately CCD sensors have been added to the focal plane of the telescope allowing both auto-targeting and the automated measurement of residual target offset. All this is implemented in embedded software.

Also, many modern theodolites, costing up to $10,000 apiece, are equipped with integrated electro-optical distance measuring devices, generally infrared based, allowing the measurement in one go of complete three-dimensional vectors — albeit in instrument-defined polar co-ordinates, which can then be transformed to a pre-existing co-ordinate system in the area by means of a sufficient number of control points. This technique is called a resection solution or free station position surveying and is widely used in mapping surveying. The instruments, "intelligent" theodolites called self-registering tacheometers or "total stations", perform the necessary operations, saving data into internal registering units, or into external data storage devices. Typically, ruggedized laptops or PDAs are used as data collectors for this purpose.

Gyrotheodolites

A gyrotheodolite is used when the north-south reference bearing of the meridian is required in the absence of astronomical star sights. This occurs mainly in the underground mining industry and in tunnel engineering. For example, where a conduit must pass under a river, a vertical shaft on each side of the river might be connected by a horizontal tunnel. A gyrotheodolite can be operated at the surface and then again at the foot of the shafts to identify the directions needed to tunnel between the base of the two shafts. Unlike an artificial horizon or inertial navigation system, a gyrotheodolite cannot be relocated while it is operating. It must be restarted again at each site.

The gyrotheodolite comprises a normal theodolite with an attachment that contains a gyroscope mounted so as to sense rotation of the Earth and from that the alignment of the meridian. The meridian is the plane that contains both the axis of the Earth's rotation and the observer. The intersection of the meridian plane with the horizontal contains the true north-south geographic reference bearing required. The gyrotheodolite is usually referred to as being able to determine or find true north.

A gyrotheodolite will function at the equator and in both the northern and southern hemispheres. The meridian is undefined at the geographic poles. A gyrotheodolite cannot be used at the poles where the Earth's axis is precisely perpendicular to the horizontal axis of the spinner, indeed it is not normally used within about 15 degrees of the pole because the east-west component of the Earth's rotation is insufficient to obtain reliable results. When available, astronomical star sights are able to give the meridian bearing to better than one hundred times the accuracy of the gyrotheodolite. Where this extra precision is not required, the gyrotheodolite is able to produce a result quickly without the need for night observations.

4

Electromagnetic Distance Measurement

EDM units operate on the principle of transmitting electromagnetic waves from an instrument to a retro-reflector, which instantly returns them to the transmitting instrument.

The instrument measures the time taken for the waves to travel this double path. Then using distance=velocity (of light) x time, the distance between instrument and prism can be obtained.

Types of EDM

There is a wide range of EDM currently available and it is beyond the scope of this manual to describe them. For simplicity EDM can be split into two types:

(a) Microwave

(b) Infra-red

The former type is required for very long distances, typically up to 100 km, and will not be considered further.

Infra-red EDM is the commonest type used for setting out purposes.

EDM can be mounted a variety of ways, either on its own tripod, or more commonly on a theodolite.

All will measure slope distances but depending on model they could be made to compute other setting out data, e.g. horizontal distances, differences, differences in height, co-ordinate differences, etc., by inputting vertical and/or horizontal angles. Further, some EDM will permit corrections to be made automatically for scale factor or prevailing atmospheric conditions whilst with others it is necessary to correct measured distances manually after the measuring process.

Depending on the type of EDM used, electronic calculators could be attached to compute additional setting out data, or electronic data loggers for automatic recording of survey data.

Use of EDM

For the actual mode of operation for individual EDM it is necessary to consult the manufacturers' handbooks.

(a) Measuring distances

i) An EDM is set over one point and a prism over the other.

ii) The theodolite telescope is aligned on the appropriate aiming mark and checks made to ensure a return signal is obtained. This is done by use of visual scales or audio tone.

Failure to achieve a return signal could be due to:

- Insufficient charge in battery
- EDM not pointing at reflector
- Reflector not pointing at EDM
- Obstruction between EDM and reflector

iii) Operate instrument to measure the distance.

iv) Some EDM have sensors to determine the slope of the line, whilst others require a vertical angle to be input, to obtain a horizontal distance.

Basic EDM measure the slope distance only. Record the vertical angle and compute the horizontal distance from – slope distance x cosine of vertical angle.

b) Measuring difference in height

(h) Proceed as above to measure the slope distance

- Again some EDM will automatically compute the difference in height at the turn of a switch, others will require the vertical angle to be input.
- With basic EDM the difference in height is calculated from – slope distance x sine of vertical angle.

c) Setting out distances

i) Mount the prism on a plumbing pole and align by theodolite. At a trial distance, measure a slope distance. Convey this information to the chainman by hand signal or radio. Then move the prism to a new trial position.

When the prism is within 2-3 m of its required position, compute a horizontal distance. Measure the difference between trial and required position with a tape and move the prism to the new position.

The distance can be checked and the ground marked as required.

ii) A "tracking" facility on some EDM provides automatic e-measurement of the distance every few seconds. The prism is set on line and the distance measured in the normal way. The EDM can then be set to "tracking" and the prism moved along the required line. The distance will be updated every few seconds. Continue until the prism is within 1-2 m of its required position and then proceed as before in (i).

iii) Some EDM have flashing coloured lights which help the chainman to maintain the correct line, whilst others have one way voice communication to the prism.

iv) The Kern remote receiver fitted to a prism is most useful as it has a display of horizontal and slope distance at the prism end. Further if the co-ordinates of the point to be set out are stored at the EDM (using HP 41CV calculator), the co-ordinates of the prism can be computed using distance and bearing and compared with the required values. The calculator programme will determine a distance that the prism should be moved towards or away from the instrument and an offset to left o right of the line. This information being conveyed to the remote receiver on the prism.

Sources of Error

Scale Error: This can be caused by variations in the modulation frequency of the EDM and is therefore proportional to the distance measured. It can only be detected by comparing the EDM distance against a known accurate standard or by laboratory test.

Scale error can also be caused by variations in the speed of light due to changing atmospheric conditions. 'Met' corrections can be applied using formulae, special calculators or graphs similar to that below for the Kern DM501.

Zero Error: This error occurs if thee are differences in the mechanical, electrical and optical centres of the EDM instrument and reflector. There is a simple field check to determine this, error for a particular EDM/prism combination. Using different prisms may result in different zero errors.

Cyclic Error: This is caused by unwanted interference between electrical signals generated in the EDM unit and can be investigated by measuring a series of known distances spread over the measuring wavelength of the instrument.

The effect of this can usually be ignored for ordinary engineering surveys but many have significance on longer lines or where a high precision is required.

Automatic Deformation Monitoring System

An automatic deformation monitoring system is a group of interacting, interrelated, or interdependent software and hardware elements forming a complex whole for deformation monitoring that, once set up, does not require human input to function.

Automatic deformation monitoring systems provide a critical function for the customer.

In many cases an automatic deformation monitoring system saved lives and prevented the loss of millions of dollars in infrastructure and income.

Automatic Deformation Monitoring System Components

Automatic deformation monitoring systems include:

- Consulting
- Sensors
- Communication
- Data acquisition software and data management
- Deformation analysis

Consulting

The consulting of the automatic deformation monitoring system covers a number of service activities that range from the first on site visit to sound out the situation and collect the requirements to the detailed project engineering with the selection of a suitable combination of measuring devices, mounting, power, communications, data centre location and data acquisition software to the installation, operation and maintenance of the system.

Sensors

Figure: *A standard geodetic monitoring instrument in the Freeport open pit mine, Indonesia*

Figure: *GNSS reference station antenna for structural monitoring of the Jiangying Bridge*

To cover all applications an automatic deformation monitoring system must support any geodetic and geotechnical measuring device (sensor) that is required by the monitoring application.

- Geodetic measuring devices measure georeferenced displacements or movements in one, two or three dimensions.

It includes the use of instruments such as total stations, levels and global navigation satellite system receivers.

- Geotechnical measuring devices measure non-georeferenced displacements or movements and related environmental effects or conditions. It includes the use of instruments such as extensometers, piezometers, rain gauges, thermometers, barometers, tilt metres, accelerometers, seismometers etc.
- Other techniques e.g. radar measuring devices.

Communication

Between measuring devices and the data acquisition software a broad range of communication alternatives are possible depending range, data rate and cost.

- Transmission cable (RS232, RS485, fibre optics)
- Local area network (LAN)
- Wireless LAN (WLAN)
- Mobile communication (GSM, GPRS, UMTS)
- WiMax

Data Acquisition Software Including Data Management

The software is the heart of the monitoring system because of its key roles in acquiring data from the attached sensors, computing meaningful values from the measurements, recording results, visualising the changes and alarming responsible persons should threshold value be exceeded.

Whilst the software is the heart of the system, the operator is always the brains because they have the sufficient skills and expertise to make considered decisions on the appropriate response to the movement, e.g. independent verification though on-site inspections, re-active controls such as structural repairs and emergency responses such as shut down processes, containment processes and site evacuation.

Deformation Analysis

Deformation analysis is concerned with determining of a measured displacement is statistically significant. The analysis can be done visually through the use of time line, scatter, vector and other plots and numerically. Numerical deformation analysis is directly related to the science of network adjustment.

Deformation Monitoring

Deformation monitoring (also referred to as Deformation survey) is the systematic measurement and tracking of the alteration in the shape or dimensions of an object as a result of stresses induced by applied loads. Deformation monitoring is a major component of logging measured values that may be used to for further computation, deformation analysis, predictive maintenance and alarming.

Deformation monitoring is primarily related to the field of applied surveying, but may be also related to the civil engineering, mechanical engineering, plant construction, soil and rock sta-bility mechanics.

The causes for deformation monitoring are changes in the bedrock, increase or decrease of weight, changes of the material properties or outside influences. The used measuring devices (1) for a deformation monitoring depend on the application (2), the chosen method (3) and the required regularity (4).

Measuring Devices

Measuring devices (or sensors) can be sorted in two main groups, geodetic and geotechnical sensors. Both measuring devices can be seamlessly combined in modern deformation monitoring.

- Geodetic measuring devices measure georeferenced displacements or movements in one, two or three dimensions. It includes the use of instruments such as total stations, levels and global navigation satellite system receivers.
- Geotechnical measuring devices measure non-georeferenced displacements or movements and related environmental effects or conditions. It includes the use of instruments such as extensometers, piezometers, rain gauges, thermometers, barometers, tilt metres, accelerometers, seismometers etc.
- Other techniques e.g. radar measuring devices.

Application

Deformation monitoring can be required for the following applications:

- Dams
- Roads
- Tunnels
- Bridges and Viaducts

- High-rise and historical buildings
- Foundations
- Construction sites
- Mining
- Landslide and Volcanoes Slopes
- Settlement areas
- Earthquake areas

Methods

Deformation monitoring can be made manually or automatically.

- Manual deformation monitoring is the operation of sensors or instruments by hand for the purpose of deformation monitoring.
- An automatic deformation monitoring system is a group of interacting, interrelated, or interdependent software and hardware elements forming a complex whole for deformation monitoring that, once set up, does not require human input to function.

Note that deformation analysis and interpretation of the data collected by the monitoring system is not included in this definition. An automatic monitoring system may be used for periodic or continuous monitoring.

Regularity and Scheduling

The monitoring regularity and time interval of the measurements must be considered depending on the application and object to be monitored. Objects can undergo both rapid, high frequency movement and slow, gradual movement. For example, a bridge might oscillates with a period of a few seconds due to the influence of traffic and wind and also be shifting gradually due to tectonic changes.

- Regularity: ranges from a days, weeks or years for manual monitoring and continuous for automatic monitoring systems.
- Measurement interval: ranges from fractions of a second to hours.

Risk Management

Deformation monitoring systems provide a proactive control of a hazard related to possible change or failure of a structure. Policyholders can reduce risk exposure before and during construction and throughout the lifecycle of the structure and hence decrease the insurance premium.

3D data Acquisition and Object Reconstruction

3D data acquisition and reconstruction is the generation of three dimensional or spatiotemporal models from sensor data. The techniques and theories, generally speaking, work with most or all sensor types including optical, acoustic, laser scanning, radar, thermal, seismic.

Acquisition

Acquisition can occur from a multitude of methods including 2D images, acquired sensor data and on site sensors.

Acquisition From 2D Images

3D data acquisition and object reconstruction can be performed using stereo image pairs. Stereo photogrammetry or photogrammetry based on a block of overlapped images is the primary approach for 3D mapping and object reconstruction using 2D images. Close-range photogrammetry has also matured to the level where cameras or digital cameras can be used to capture the close-look images of objects, e.g., buildings, and reconstruct them using the very same theory as the aerial photogrammetry. An example of software which could do this is VexcelFotoG 5. This software has now been replaced by Vexcel GeoSynth. Another similar software programme is Microsoft Photosynth.

A semi-automatic method for acquiring 3D topologically structured data from 2D aerial stereo images has been presented by Sisi Zlatanova. The process involves the manual digitizing of a number of points necessary for automatically reconstructing the 3D objects. Each reconstructed object is validated by superimposition of its wire frame graphics in the stereo model. The topologically structured 3D data is stored in a database and are also used for visualization of the objects. Software used for 3D data acquisition using 2D images include e.g. ENSAIS Engineering College TIPHON (Traitement d'Image et PHOtogrammétrie Numérique). CyberCity 3D Modeler, ORPHEUS, ...

A method for semi-automatic building extraction together with a concept for storing building models alongside terrain and other topographic data in a topographical information system has been developed by Franz Rottensteiner. His approach was based on the integration of building parameter estimations into the photogrammetry process applying a hybrid modelling scheme. Buildings are decomposed into a set of simple primitives that are reconstructed individually and are then combined by Boolean operators. The internal data structure of both the primitives and the compound building models are based on the boundary representation methods

Multiple images are used in Zeng's approach to surface reconstruction from multiple images. A central idea is to explore the integration of both 3D stereo data and 2D calibrated images. This approach is motivated by the fact that only robust and accurate feature points that survived the geometry scrutiny of multiple images are reconstructed in space.

The density insufficiency and the inevitable holes in the stereo data should then be filled in by using information from multiple images. The idea is thus to first construct small surface patches from stereo points, then to progressively propagate only reliable patches in their neighborhood from images into the whole surface using a best-first strategy. The problem thus reduces to searching for an optimal local surface patch going through a given set of stereo points from images.

Multi-spectral images are also used for 3D building detection. The first and last pulse data and the normalized difference vegetation index are used in the process.

New measurement techniques are also employed to obtain measurements of and between objects from single images by using the projection, or the shadow as well as their combination. This technology is gaining attention given its fast processing time, and far lower cost than stereo measurements. GeoTangoSilverEye technology is the first of this kind commercial product that can produce very realistic city models and buildings from single satellite and aerial images.

Acquisition From Acquired Sensor Data

Semi-Automatic building extraction from LIDAR Data and High-Resolution Images is also a possibility. Again, this approach allows modelling without physically moving towards the location or object. From airborne LIDAR data, digital surface model (DSM) can be generated and then the objects higher than the ground are automatically detected from the DSM. Based on general knowledge about buildings, geometric characteristics such as size, height and shape information are then used to separate the buildings from other objects.

The extracted building outlines are then simplified using an orthogonal algorithm to obtain better cartographic quality. Watershed analysis can be conducted to extract the ridgelines of building roofs. The ridgelines as well as slope information are used to classify the buildings per type. The buildings are then reconstructed using three parametric building models (flat, gabled, hipped).

Acquisition From on-site Sensors

LIDAR and other terrestrial laser scanning technology offers the fastest, automated way to collect height or distance information. LIDAR or laser for height measurement of buildings is becoming very promising. Commercial applications of both airborne LIDAR and ground laser scanning technology have proven to be fast and accurate methods for building height extraction. The building extraction task is needed to determine building locations, ground elevation, orientations, building size, rooftop heights, etc. Most buildings are described to sufficient details in terms of general polyhedra, i.e., their boundaries can be represented by a set of planar surfaces and straight lines. Further processing such as expressing building footprints as polygons is used for data storing in GIS databases.

Using laser scans and images taken from ground level and a bird's-eye perspective, Fruh and Zakhor present an approach to automatically create textured 3D city models. This approach involves registering and merging the detailed facade models with a complementary airborne model. The airborne modelling process generates a half-metre resolution model with a bird's-eye view of the entire area, containing terrain profile and building tops. Ground-based modelling process results in a detailed model of the building facades. Using the DSM obtained from airborne laser scans, they localize the acquisition vehicle and register the ground-based facades to the airborne model by means of Monte Carlo localization (MCL). Finally, the two models are merged with different resolutions to obtain a 3D model.

Using an airborne laser altimeter, Haala, Brenner and Anders combined height data with the existing ground plans of buildings. The ground plans of buildings had already been acquired either in analog form by maps and plans or digitally in a 2D GIS. The project was done in order to enable an automatic data capture by the integration of these different types of information. Afterwards virtual reality city models are generated in the project by texture processing, e.g. by mapping of terrestrial images. The project demonstrated the feasibility of rapid acquisition of 3D urban GIS. Ground plans proved are another very important source of information for 3D building reconstruction.

Compared to results of automatic procedures, these ground plans proved more reliable since they contain aggregated information which has been made explicit by human interpretation. For this reason, ground plans, can considerably reduce costs in a reconstruction project. An example of existing ground plan data usable in building reconstruction is the Digital Cadastral map, which provides information on the distribution of property, including the borders of all agricultural areas and the ground plans of existing buildings. Additionally information as street names and the usage of buildings (e.g. garage, residential building, office block, industrial building, church) is provided in the form of text symbols. At the moment the Digital Cadastral map is build up as a data base covering an area, mainly composed by digitizing preexisting maps or plans.

Software

Software used for airborne laser scanning includes OPALS (Orientation and Processing of Airborne Laser Scanning data), ...

Cost

The financial costs of 3D data acquisition and object reconstruction are not to be underestimated.

- Terrestric laserscandevices (pulse or phase devices) + processing software generally start at a price of 150,000 €. Some less precise devices (as the Trimble VX) cost around 75,000€.
- Terrestric LIDAR systems cost around 300,000 €.
- Systems using regular still cameras mounted on RC helicopters (Photogrammetry) are also possible, and cost around 25,000€. Systems that use still cameras with balloons are even cheaper (around 2,500 €), but require additional manual processing. As the manual processing takes around 1 month of labour for every day of taking pictures, this is thus also still an expensive solution in the long run.
- Obtaining satellite images is also an expensive endeavour. High resolution stereo images (0.5 m resolution) cost around 11,000€. Image satellites include Quikbird, Ikonos. High resolution monoscopic images cost around 5,500€. Somewhat lower resolution images (e.g. from the CORONA satellite; with a 2m resolution) cost around 1.000€ per 2 images. Note that Google Earth images are too low in resolution to make an accurate 3D model.

Object Reconstruction

After the data has been collected, the acquired (and sometimes already processed) data from images or sensors needs to be reconstructed. This may be done in the same programme or in some cases, the 3D data needs to be exported and imported into another programme for further refining, and/or to add additional data. Such additional data could be gps-location data, ... Also, after the reconstruction, the data might be directly implemented into a local (GIS) map or a worldwide map such as Google Earth.

Software

Several software packets are used in which the acquired (and sometimes already processed) data from images or sensors is imported. The software packets include (in alphabetical order) :

- Canoma
- Cyclone
- Leica Photogrammetry Suite
- MountainsMap SEM (microscopy applications only)
- Neitra 3D pro
- Orthoware
- PhotoModeler
- SketchUp
- Smart3Dcapture (acute3D)
- Rhinophoto

5

Aerial Survey

Aerial survey is a geomatics method of collecting information by using aerial photography, LiDAR or from remote sensingimagery using other bands of the electromagnetic spectrum, such as infrared, gamma, or ultraviolet. It can also refer to the chart or map made by analysing a region from the air. This is typically done using aeroplanes, helicopters, UAVs such as the InView Unmanned Aircraft System and in history with balloons. Aerial survey should be distinguished from satellite imagery technologies because of its better resolution, quality and atmospheric conditions.

Today, aerial survey is often recognized as a synonym for aerophotogrammetry, part of photogrammetry where the camera is placed in the air. Measurements on aerial images are provided by photogrammetric technologies and methods.

Aerial surveys can provide information on many things not visible from the ground.

Terms Used in Aerial Survey

- exposure station or air station the position of the optical centre of the camera at the moment of exposure.
- flying height the eleveation of the exposure station above the datum (usually mean sea level).
- altitude the vertical distance of the aircraft above the earth surface.
- tilt the angle between the aerial camera and the horizontal axis perpendicular to the line of flight.
- tip the angle between the aerial camera and the line of flight.

- principal point the point of intersection of the optical axis of the aerial camera with the photographical plane
- isocentre the point on the areal photograph in which the bisector of the angle of tilt meets the photograph.
- nadir point the image of the nadir, i.e. the point on the areal photograph where a plumbline dropped from the front nodal point pierces the photograph.
- scale ratio of the focal length of the camera objective and the distance of the exposure station from the ground

Aerial surveys are used for:

Aerial view of the Paranal Observatory, created by the non-profit initiative Wings for Science which offers aerial support to public research organisations.

- Archaeology
- Fishery surveys
- Hydrocarbon exploration
- Land survey
- Mining
- Monitoring wildlife and insect populations, called aerial census or sampling.
- Monitoring vegetation and ground cover

- Reconnaissance
- Transportation Projects in conjunction with Ground Surveys (Roadway, Bridge, Interstate)

Aerial survey uses a measuring camera where the elements of the interior orientation are known, but a camera that has much larger focal length and film and more lenses are used.

Baseline (Surveying)

In the United States Public Land Survey System, a baseline is the principal east-west line upon which all rectangular surveys in a defined area are based. The baseline meets its corresponding principal meridian at the point of origin, or *initial point*, for the land survey. For example, the baseline for Nebraska and Kansas is shared as the border for both states, at the 40th parallel north.

More specifically a baseline may be the line that divides a survey township between north and south.

Often, a baseline is marked by other features such as a road or boundary between counties.

"Baseline Road" in the United States

Many communities in the United States have roads that run along survey baselines, many of which are named to reflect that fact. Some examples:

- In Little Rock, Arkansas, Baseline Road follows the baseline used by surveyors of the Louisiana Purchase.
- In Colorado, Baseline Road in Boulder marks the 40th parallel, or the western extension of the Kansas-Nebraska boundary, which is also the boundary between Adams and Weld counties.
- In Arizona, the baseline near the Phoenix metro area is marked by Baseline Road.
- In Southern California, from Highland to San Dimas, the baseline is marked by Baseline Road.
- In Michigan, the baseline for the Michigan Survey forms the boundary between the second and third tiers of counties and in many portions, discontinuous segments of road along the baseline are known as "Baseline Road." 8 Mile Road in the Detroit area runs along the Michigan Baseline and was formerly known as "Baseline Road."

- Baseline Road in Hillsboro, Oregon, generally follows the Willamette Baseline which intersects the Willamette Meridian at the Willamette Stone State Park.

Canada

In Canadian land surveying, a base line is one of the many principal east-west lines that correspond to 4 tiers of townships (2 tiers north and 2 south). The base lines are about 24 miles apart, with the first base line at the 49th parallel, the Western US-Canadian border. It is therefore equivalent to the *standard parallel* in the US system. In Ontario, a baseline forms a straight line parallel a geographical feature (mostly a lake, especially Lake Ontario or Lake Erie) that serves as a reference line for surveying a grid of property lots. The result of this surveying is the concession road and sideline system in use today.

Benchmark (Surveying)

The term bench mark, or benchmark, originates from the chiseled horizontal marks that surveyors made in stone structures, into which an angle-iron could be placed to form a "bench" for a levelling rod, thus ensuring that a levelling rod could be accurately repositioned in the same place in the future. These marks were usually indicated with a chiselled arrow below the horizontal line.

The term is generally applied to any item used to mark a point as an elevation reference. Frequently, bronze or aluminium disks are set in stone or concrete, or on rods driven deeply into the earth to provide a stable elevation point.

The height of a benchmark is calculated relative to the heights of nearby benchmarks in a network extending from a *fundamental benchmark*. A fundamental benchmark is a point with a precisely known relationship to the level datum of the area, typically mean sea level. The position and height of each benchmark is shown on large-scale maps.

The terms "height" and "elevation" are often used interchangeably, but in many jurisdictions they have specific meanings; "height" commonly refers to a local or relative difference in the vertical (such as the height of a building), whereas "elevation" refers to the difference from a nominated reference surface (such as sea-level, or a mathematical/geodetic model that approximates the sea level known as the geoid). Elevation may be specified as normal height (above a reference ellipsoid), orthometric height, or dynamic height which have slightly different definitions.

Other Types of Survey Marks

Figure: *An Ordnance Survey flush bracket*

Triangulation points, also known as trig points, are marks with a precisely established horizontal position. These points may be marked by disks similar to benchmark disks, but set horizontally, and are also sometimes used as elevation benchmarks. Prominent features on buildings such as the tip of a churchspire or a chimney stack are also used as reference points for triangulation.

In the United Kingdom, triangulation points are often set in large concrete markers, which as well as functioning as a triangulation point, have a benchmark set into the side. With the increasing use of GPS and electronic distance measuring devices, the same techniques and equipment are used to fix the horizontal and vertical position of a survey marker at the same moment, and therefore the marks are usually regarded as "fixed in three dimensions".

Agencies Responsible for Benchmarks

Benchmarks are typically placed ("monumented") by a government agency or private survey firm, and many governments maintain a register of these marks so that the records are available to all. These records are usually in the form of a geographically searchable database (computer or map-based), with links to sketches, diagrams, photos of the marks, and any other technical details.

Government agencies that place and maintain records of benchmarks include:

- Canada
  - Natural Resources CanadaGeodetic Survey Division (History of the Geodetic Survey Division)
- France
  - Institut Géographique National (IGN on Wiki FR)
- Italy
  - Istituto Geografico Militare - Servizio Geodetico
- Japan
  - Geographical Survey Institute (GSI)
- New Zealand
  - Land Information New Zealand
- Spain
  - Instituto Geográfico Nacional (IGN)
- United Kingdom
  - Ordnance Survey
- United States
  - The National Geodetic Survey (NGS; formerly U.S. Coast & Geodetic Survey)
  - The United States Geological Survey (USGS)
  - The United States Army Corps of Engineers (USACE)

Bundle Adjustment

Given a set of images depicting a number of 3D points from different viewpoints, bundle adjustment can be defined as the problem of simultaneously refining the 3D coordinates describing the scene geometry as well as the parameters of the relative motion and the optical characteristics of the camera(s) employed to acquire the images, according to an optimality criterion involving the corresponding image projections of all points.

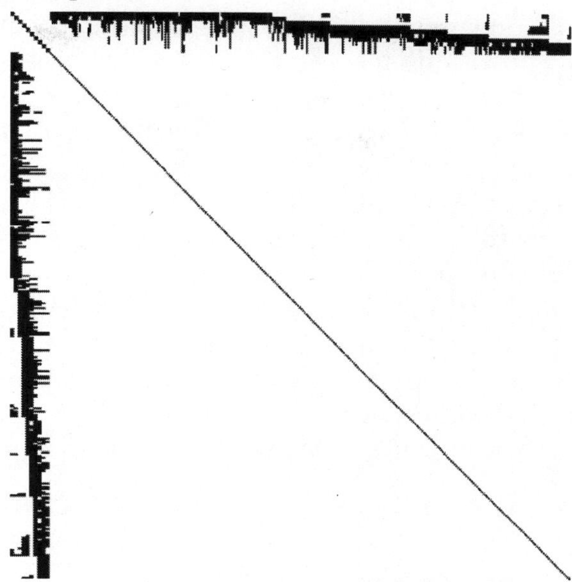

Figure: *A sparse matrix obtained when solving a modestly sized bundle adjustment problem. This is the sparsity pattern of a 992×992 normal equations (i.e. approximate Hessian) matrix. Black regions correspond to nonzero blocks.*

Bundle adjustment is almost always used as the last step of every feature-based 3D reconstruction algorithm. It amounts to an optimization problem on the 3D structure and viewing parameters (i.e., camera pose and possibly intrinsic calibration and radial distortion), to obtain a reconstruction which is optimal under certain assumptions regarding the noise pertaining to the observed image features: If the image error is zero-mean Gaussian, then bundle adjustment is the Maximum Likelihood Estimator. Its name refers to the bundles of light rays originating from each 3D feature and converging on each camera's optical centre, which are adjusted optimally with respect to both the structure and viewing parameters (similarity in meaning to categoricalbundle seems a pure coincidence). Bundle adjustment was originally conceived in the field of photogrammetry during 1950s and has increasingly been used by computer vision researchers during recent years.

Bundle adjustment boils down to minimizing the reprojection error between the image locations of observed and predicted image points, which is expressed as the sum of squares of a large number of nonlinear, real-valued functions. Thus, the minimization is achieved using nonlinear least-squares algorithms. Of these, Levenberg–Marquardt has proven to be one of the most successful due to its ease of implementation and its use of an effective damping strategy that lends it the ability to converge quickly from a wide range of initial guesses. By iteratively linearizing the function to be minimized in the neighborhood of the current estimate, the Levenberg–Marquardt algorithm involves the solution of linear systems known as the normal equations. When solving the minimization problems arising in the framework of bundle adjustment, the normal equations have a sparse block structure owing to the lack of interaction among parameters for different 3D points and cameras. This can be exploited to gain tremendous computational benefits by employing a sparse variant of the Levenberg–Marquardt algorithm which explicitly takes advantage of the normal equations zeros pattern, avoiding storing and operating on zero elements.

6

Environmental Surveying

Environmental surveying is the title of a profession within the wider field of surveying, the practitioners of which are known as environmental surveyors.

Environmental surveyors use surveying techniques to understand the potential impact of environmental factors on real estate and construction developments, and conversely the impact that real estate and construction developments will have on the environment.

Professional Activities

The exact activities that make up the day to day work of an environmental surveyor vary from surveyor to surveyor and from project to project. Two environmental surveyors could have careers that consist of quite different professional activities depending on their and their practices area of specialisation.

In the strictest sense, the field of environmental surveying is distinct from that of environmental consultancy. Environmental consultancies may have some overlap with the work of environmental surveyors, but may be members of different professional bodies and may carry out activities not involving the built environment. They may for example be involved with arboriculture the specifics of which fall out of the remit of environmental surveyors.

The terms are however sometimes used interchangeably, and practices often use the term consultants if the practice is seeking a wider client base than would be attracted to a pure Environmental Surveyor practice.

Main Areas of Operation

The main areas of operation for environmental surveyors in the UK include:

- Flood risk assessment- This is to assess how likely it is that a building or proposed building will flood. If a building is thought to be at risk it will receive a designation of either Band 1 (200:1 chance of flooding in a year) Band 2 (between 200:1 and 75:1 chance of flooding annually) or Band 3 (greater than a 75:1 chance of flooding annually, currently thought to account for around 4% of flood risk properties in the UK).
- Contaminated land assessment- Contaminated land surveys are carried out to assess the level of threat posed to existing or proposed buildings. Land can be contaminated if it is on or near a site that is currently or has in the past been used for industrial or waste disposal purposes. Such surveys form part of the due diligence that must be carried out before construction or modification of a real estate asset can begin. Both during and after construction a contaminated land survey could be an important factor in informing risk management strategies.
- Environmental screenings- Provide a general overview of environmental risks proposed to an existing or proposed real estate development. The screening can help gain a picture of: whether or not the property in question might have been damaged by undermining, whether the property might be susceptible to ground gas, the closeness of government licensed waste disposal facilities and an assessment of a properties water resource vulnerability to contamination.
- Fire risk assessment- All work premises in the UK must have a fire risk assessment. The assessment is designed to ascertain what could start a fire, how the fire could be dealt with and ensuring that the staff will be sufficiently warned of a fire, have exits from the building and a safe place to congregate afterwards.
- Asbestos surveys- Because asbestos is extremely dangerous material to the health of humans, its use is strictly controlled. 52 countries globally have now banned the substance. The substance is banned by the European Union, with the exception of its use in a very limited number of specific industrial applications. Because of its widespread use in the building

industry before banning, many existing buildings contain asbestos and sites where buildings have been previously may have been contaminated with it. For this reason buildings may need an asbestos survey to ascertain the level of use of the substance and the level of contamination to the site this has resulted in.

Techniques

Environmental Surveyors use a range of techniques to assess the environmental conditions of an area and compile their reports.

- Historical data is drawn from maps and older survey information to establish the exact boundaries of a property, and are also used to see if there has been any historical pollution or waste dumping on the site.
- Water Sampling allows Environmental Surveyors to gain a picture of the quality of and pollution levels in local water sources.
- In a similar way to Water Sampling, Earth Sampling can be used to analyse the level of pollutants in an area's soil.
- Geometric data may used to establish areas that are likely to flood or monitor the spread of pollutants.
- Geographic information systems (GISs) can cross reference map data with statistical data. If an Environmental Surveyor was compiling a flood report for a building and wanted to establish the odds of a property flooding in any given year then they could cross reference the geographic location of a property with historically obtained statistical data on flooding in the area.
- Visual Inspection might be used if for example the surveyor wished to establish the level of asbestos contamination to a given property. This might be enhanced by or presented in reference to the collection

Chartered Practitioners

In the UK as well as in many other countries globally, recognition by the Royal Institution of Chartered Surveyors (RICS) is looked upon as conferring a high professional standard, and guaranteeing a level of quality in the work of its member surveyors. Environmental Surveyors form one professional group within RICS and are listed in their Land Professional Group. To achieve the status of chartered environmental surveyor, the candidate must pass an assessment of professional competencies (APC). This consists of completing structured work experience and providing written documents as evidence of the activities carried out during this work experience. Finally the candidate must pass an hour long oral exam.

All surveyors regardless of their field are required to demonstrate mastery of RICS core competencies, and then move on to demonstrate knowledge of competencies in their specific fields. Competencies specific to environmental surveying include:

- sustainability
- contaminated land
- environmental assessment
- environmental audit
- laboratory procedures
- management of the natural environment and landscape

Outside of the UK, other professional bodies may offer equivalent designations to signify the professional level of environmental surveyors.

Hansen's Problem

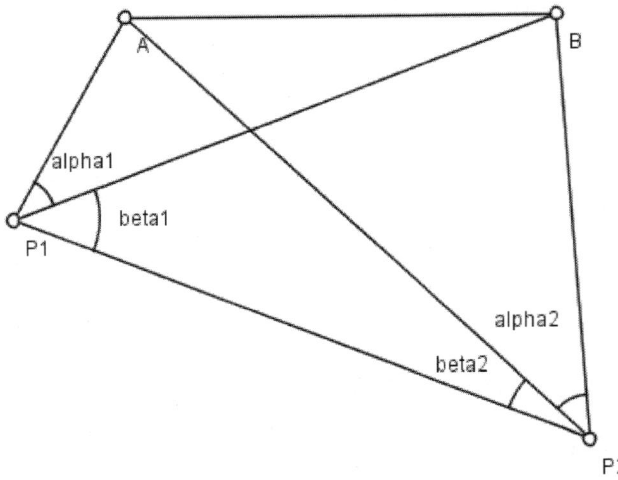

Hansen's problem is a problem in planar surveying, named after the astronomer Peter Andreas Hansen (1795–1874), who worked on the geodetic survey of Denmark.

There are two known points A and B, and two unknown points P_1 and P_2. From P_1 and P_2 an observer measures the angles made by the lines of sight to each of the other three points. The problem is to find the positions of P_1 and P_2.

Since it involves observations of angles made at unknown points, the problem is an example of resection (as opposed to intersection).

Solution Method Overview

Define the Following Angles: $\gamma = P_1AP_2$, $\delta = P_1BP_2$, $\varphi = P_2AB$, $\psi = P_1BA$. As a first step we will solve for φ and ψ. The sum of these two unknown angles is equal to the sum of β_1 and β_2, yielding the following equation:

$$\phi + \psi = \beta_1 + \beta_2$$

A second equation can be found more laboriously, as follows. The law of sines yields

$$\frac{AB}{P_2B} = \frac{\sin \alpha_2}{\sin \phi} \text{ and}$$

$$\frac{P_2B}{P_1P_2} = \frac{\sin \beta_1}{\sin \delta}$$

combining these together we get

$$\frac{AB}{P_1P_2} = \frac{\sin \alpha_2 \sin \beta_1}{\sin \phi \sin \delta}$$

An entirely analogous reasoning on the other side yields

$$\frac{AB}{P_1P_2} = \frac{\sin \alpha_1 \sin \beta_2}{\sin \psi \sin \gamma}$$

Setting these two equal gives

$$\frac{\sin \phi}{\sin \psi} = \frac{\sin \gamma \sin \alpha_2 \sin \beta_1}{\sin \delta \sin \alpha_1 \sin \beta_2} = k$$

Using a known trigonometric identity this ratio of sines can be expressed as the tangent of an angle difference:

$$\tan \frac{\phi - \psi}{2} = \frac{k-1}{k+1} \tan \frac{\phi + \psi}{2}$$

This is the second equation we need. Once we solve the two equations for the two unknowns ϕ and ψ, we can use either of the two expressions above for $\frac{AB}{P_1P_2}$ to find P_1P_2 since AB is known. We can then find all the other segments using the law of sines.

Solution Algorithm

We are given four angles (α_1, β_1, α_2, β_2) and the distance AB. The calculation proceeds as follows:

- Calculate $\gamma = \pi - \alpha_1 - \beta_1 - \beta_2$, $\delta = \pi - \alpha_2 - \beta_1 - \beta_2$
- Calculate $k = \dfrac{\sin\gamma \sin\alpha_2 \sin\beta_1}{\sin\delta \sin\alpha_1 \sin\beta_2}$
- Let $s = \beta_1 + \beta_2$, $d = 2\arctan\left[\dfrac{k-1}{k+1}\tan(s/2)\right]$ and then $\phi = (s+d)/2$, $\psi = (s-d)/2$.
- Calculate

  $$P_1P_2 = AB\frac{\sin\phi \sin\delta}{\sin\alpha_2 \sin\beta_1},$$

  or equivalently,

  $$P_1P_2 = AB\frac{\sin\psi \sin\gamma}{\sin\alpha_1 \sin\beta_2}.$$

If one of these fractions has a denominator close to zero, use the other one.

Heliotrope (Instrument)

The heliotrope is an instrument that uses a mirror to reflect sunlight over great distances to mark the positions of participants in a land survey. The heliotrope was invented in 1821 by the German mathematician Carl Friedrich Gauss. The word "heliotrope" is taken from the Greek: *helios*, meaning "sun", and *tropos*, meaning "turn". It is a fitting name for an instrument which can be turned to reflect the sun toward a given point.

The heliotrope was utilized by surveyors as a specialized form of target; it was employed during large triangulation surveys where, because of the great distance between stations (usually twenty miles or more), a regular target would appear indistinct. Heliotropes have been used repeatedly as survey targets at ranges of over 100 miles. In California, in 1878, a heliotrope on Mount Saint Helena was surveyed by B.A. Colonna of the USCGS from Mount Shasta, a distance of 192 miles (309 km).

The heliotrope was limited to use on sunny days and was further limited (in regions of high temperatures) to mornings and afternoons when atmospheric aberration least affected the instrument-man's line of sight. The heliotrope operator was called a "heliotroper" or "flasher" and would sometimes employ a second mirror for communicating with the instrument station through heliography, a signalling system using impulsed reflecting surfaces. The inventor of the Heliograph, a similar instrument specialized for signalling, was inspired by observing the use of heliotropes in the survey of India.

Hydrographic Survey

Hydrographic survey is the science of measurement and description of features which affect maritime navigation, marine construction, dredging, offshore oil exploration/drilling and related activities. Strong emphasis is placed on soundings, shorelines, tides, currents, sea floor and submerged obstructions that relate to the previously mentioned activities. The term Hydrography is sometimes used synonymously to describe Maritime Cartography, which in the final stages of the hydrographic process uses the raw data collected through hydrographic survey into information usable by the end user.

Hydrography is collected under rules which vary depending on the acceptance authority. Traditionally conducted by ships with a sounding line or echo sounding, surveys are increasingly conducted with the aid of aircraft and sophisticated electronic sensor systems in shallow waters.

National and International Maritime Hydrography

Hydrographic offices evolved from naval heritage and are usually found within national naval structures, for example Spain's Instituto Hidrográfico de la Marina. Coordination of those organizations and product standardization is voluntarily joined with the goal of improving hydrography and safe navigation is conducted by the International Hydrographic Organization (IHO). The IHO publishes Standards and Specifications followed by member states as well as Memoranda of Understanding and Co-operative Agreements with hydrographic survey interests.

The product of such hydrography is most often seen on nautical charts published by the national agencies and required by the International Maritime Organization (IMO), the Safety of Life at Sea (SOLAS) and national regulations to be carried on vessels for safety purposes. Increasingly those charts are provided and used in electronic form unders IHO standards.

History and Responsibilities

The United Kingdom has a long hydrographic history officially begun with the 1683 appointment of Captain Grenville Collins as Hydrographer to the King. With the Royal Navy dominating the seas hydrography grew to a worldwide hydrographic activity. That tradition extended to the nations with a common legacy in the Empire, for example, the Australian Hydrographic Service. The British Admiralty Hydrographic Office became the United Kingdom Hydrographic Office which continues the legacy within the Ministry of Defence with responsibility for the Admiralty Charts. The Royal Navy maintains a number of hydrographic survey vessels to continue the work today.

Argentina

The Argentine Hydrographic Service was established in 1879.

Australia

Hydrographic services are provided by the Royal Australian Navy Hydrographic Service.

Canada

Hydrographic services are provided by the Canadian Hydrographic Service.

France

Hydrographic services are provided by the Naval Hydrographic and Oceanographic Service.

Germany

Hydrographic services are provided by the Federal Maritime and Hydrographic Agency.

The Netherlands

The Dutch "Dienst der Hydrografie" is part of the Royal Netherlands Navy, since 8 January 1488.

United States

In United States statutory authority for hydrographic surveys of territorial waters and the Exclusive Economic Zone (EEZ) lies with the National Oceanic and Atmospheric Administration (NOAA). NOAA hydrographic surveys are conducted by the National Ocean Service, a uniformed corps within NOAA and a fleet of survey vessels based at two major centres. The organic survey assets are supplemented by other agencies and contract surveys in order to survey the large areas within its responsibility. Those were identified in the NOAA Hydrographic Survey Priorities (NHSP) - East Coast alone as being 3,603 square miles (9,330 km^2) classified as critical. The 2009 status shows 29,412 square nautical miles (100,900 km^2) out of 510,841 square nautical miles (1,752,000 km^2) "Navigationally Significant" were completed. The NOAA Office of Coast Survey, Hydrographic Surveys Division estimates it has awarded approximately $250 million in contracts for hydrographic surveying and related support since 1994.

For inland surface waters such as rivers, streams and inland lakes the U.S. Geological Survey (USGS) has national responsibility. USGS coordinates survey data collection and publishes a National Hydrography Dataset that is designed to be used with geographic information systems (GIS). Other federal agencies such as the Environmental Protection Agency and the U.S. Fish and Wildlife Service use these data and, along with state and local hydrographic collection organizations, contribute to the national hydrographic data base. The Environmental Protection Agency conducts or contracts for surveys on projects such as the GE/Hudson River Super Fund site.

The U.S. Coast Guard conducts hydrographic survey operations, particularly in the Polar regions.

The National Geospatial-Intelligence Agency (NGA) oversees charting of international waters for Department of Defence purposes. The Navy's Naval Oceanographic Office conducts many the oceanic surveys. The U.S. Army Corps of Engineers conducts hydrographic surveys supporting its responsibility for the major waterway projects that include navigation and flood control. Hydrographic data from those surveys is published by districts. Such data is incorporated into both NOAA and NGA products and the Corps engages in efforts to improve hydrographic collection methods. Military combat organizations such as the Navy's SEAL and engineering units have specialized hydrographic reconnaissance survey capability.

The NOAA Office of Coast Survey, Coast Survey Partners web page offers a useful list and summary of major player activities, government and private, with links to those partner web sites.

Hydrographic Survey Conducted by Non-national Agencies

Governmental entities below national level conduct or contract for hydrographic surveys for waters within their jurisdiction with both internal and contract assets. Such surveys are commonly conducted by or under the standards approved by or the supervision of national organizations, particularly when the use is for the purposes of chart making/distribution or dredging of state controlled waters.

In the United States there is coordination with the National Hydrography Dataset in survey collection and publication. State environmental organizations publish hydrographic data relating to their mission.

Hydrographic Survey Conducted by Private Organizations

Large scale hydrographic and geophysical survey is conducted by commercial entities, particularly in the dredging, marine construction, oil exploration & drilling industries. Industry installing submarine cable for communications or power require detailed surveys of cable routes prior to installation with increased use of acoustic imagery equipment previously found only in military applications. There are specialized companies with both the assets and expertise to contract for such surveys with both commercial and governmental entities.

Companies, Universities and investment groups will often fund Hydrographic surveys of public waterways prior to developing areas adjacent those waterways. Survey firms are also contracted to survey in support of design and engineering firms that are under contract for large public projects. Private surveys are also conducted before dredging operations and after these operations are completed. Companies with large private slips, docks or other water front installations have their facilities and the open water near their facilities surveyed regularly.

Crowd sourcing is also entering hydrographic surveying, with projects such as TeamSurv and ARGUS. Here, volunteer vessels record position, depth and time data using their standard navigation instruments, and then the data is processed on the server for speed of sound, tidal and other corrections. With this approach there is no need for a specific survey vessel, or for professionally qualified surveyors to be on board, as the expertise is in the data processing that occurs once the data is uploaded to the server after the voyage. Apart from obvious cost savings, this also gives a continuous survey of an area, but the drawbacks are time in recruiting loggers and getting a high enough density of data. Also, although accurate to 0.1 - 0.2m, this approach does not have the accuracy and coverage of a multi-beam survey.

Process

Modern surveying relies as much on software as hardware. In suitable shallow water areas Light Detection and Ranging (LIDAR) may be used. Equipment can be installed on inflatable craft, such as Zodiacs, small craft, AUVs (Autonomous Underwater Vehicles), UUVs (Unmanned Underwater Vehicles) or large ships, and can include sidescan, single beam and multibeam equipment. At one time different data collection methods and standards were used in collecting hydrographic data for maritime safety and for scientific or engineering bathymetric charts. Increasingly with aid of improved collection techniques and computer processing the data is collected under one standard and extracted for the specific use.

After data is collected, it has to undergo post-processing. A massive amount of data is collected during the typical Hydrographic survey, often several soundings per square foot. Depending on the final use (navigation charts, Digital Terrain Model, volume calculation for dredging, topography, Bathymetry) this data must be thinned out. It must also be error corrected (bad soundings,) and corrected for the effects of tides, waves/heave, water level and water temperature differences (thermoclines.) Usually the surveyor has additional data collection equipment on site to record the data required for correcting the soundings. Final output of charts can be created in a combination of speciality charting software or a CAD package, usually Autocad.

With crowd sourced surveying, although the accuracy of the individual measurements are not as accurate as with a traditional survey, the algorithms used rely on a high data density to produce final results that are more accurate than the single measurements. Comparison against multi-beam surveys indicates an accuracy of around +/- 0.1 - 0.2m.

Gyrotheodolite

A gyro-theodolite is a surveying instrument composed of a gyroscope mounted to a theodolite. It is used to determine the orientation of true north by locating the meridian direction. It is the main instrument for orientation in mine surveying and in tunnel engineering, where astronomical star sights are not visible.

History

In 1852, the French physicist Léon Foucault discovered a gyro with two degrees of freedom points north. This principle was adapted by Max Schuler in 1921 to build the first surveying gyro. In 1949, the gyro-theodolite - at that time called a "meridian pointer" or "meridian indicator" - was first used by the Clausthal Mining Academy underground. Several years later it was improved with the addition of autocollimation telescopes. In 1960, the Fennel Kassel company produced the first of the KT1 series of gyro-theodolites. Fennel Kassel and others later produced gyro attachments that can be mounted on normal theodolites.

Operation

A gyroscope is mounted in a sphere, lined with Mu-metal to reduce magnetic influence, connected by a spindle to the vertical axis of the theodolite. The battery-powered gyro wheel is rotated at 20,000 rpm or more, until it acts as a north-seeking gyroscope. A separate optical system within the attachment permits the operator to rotate the theodolite and thereby bring a zero mark on the attachment into coincidence with the gyroscope spin axis. By tracking the spin axis as it oscillates about the meridian, a record of the azimuth of a series of the extreme stationary points of that oscillation may be determined by reading the theodolite azimuth circle.

A midpoint can later be computed from these records that represents a refined estimate of the meridian. Careful setup and repeated observations can give an estimate that is within about 10 arc seconds of the true meridian. This estimate of the meridian contains errors due to the zero torque of the suspension not being aligned precisely with the true meridian and to measurement errors of the slightly damped extremes of oscillation. These errors can be moderated by refining the initial estimate of the meridian to within a few arc minutes and correctly aligning the zero torque of the suspension.

When the spinner is released from restraint with its axis of rotation aligned close to the meridian, the gyroscopic reaction of spin and Earth's rotation results in precession of the spin axis in the direction of alignment with the plane of the meridian. This is because the daily rotation of the Earth is in effect continuously tilting the east-west axis of the station. The spinner axis then accelerates towards and overshoots the meridian, it then slows to a halt at an extreme point before similarly swinging back towards the initial point of release. This oscillation in azimuth of the spinner axis about the meridian repeats with a period of a few minutes. In practice the amplitude of oscillation will only gradually reduce as energy is lost due to the minimal damping present. Gyro-theodolites employ an undamped oscillating system because a determination can be obtained in less than about 20 minutes, while the asymptotic settling of a damped gyro-compass would take many times that before any reasonable determination of meridian could possibly be made.

When not in operation, the gyroscope assembly is anchored within the instrument. The electrically powered gyroscope is started while restrained and then released for operation. During operation the gyroscope is supported within the instrument assembly, typically on a thin vertical tape that constrains the gyroscope spinner axis to remain horizontal. The alignment of the spin axis is permitted to rotate in azimuth by only the small amount required during operation. An initial approximate estimate of the meridian is needed. This might be determined with a magnetic compass, from an existing survey network or by the use of the gyro-theodolite in an extended tracking mode.

Uses

Gyro-theodolites are primarily used in the absence of astronomical star sights. For example, where a conduit must pass under a river, a vertical shaft on each side of the river might be connected by a horizontal tunnel. A gyro-theodolite can be operated at the surface and then again at the foot of the shafts to identify the directions needed to tunnel between the base of the two shafts. During the construction of the Channel Tunnel, which runs under the English Channel from France to the UK, gyro-theodolites were used to prevent and correct the tunnels from curving.

Limitations

Although a gyro-theodolite functions at the equator and in both the northern and southern hemispheres, it cannot be used at either the North Pole or South Pole, where the Earth's axis is precisely perpendicular to the horizontal axis of the spinner and the meridian is undefined. Gyro-theodolites are not normally used within about 15 degrees of the pole because the east-west component of the Earth's rotation is insufficient to obtain reliable results.

Unlike an artificial horizon or inertial navigation system, a gyro-theodolite cannot be relocated while it is operating. It must be restarted again at each site.

When available, astronomical star sights are able to give the meridian bearing to better than one hundred times the accuracy of the gyro-theodolite. Where this extra precision is not required, the gyro-theodolite is able to produce a result quickly without the need for night observations.

Initial Point

In surveying, an initial point is a datum (a specific point on the surface of the earth) that marks the beginning point for a cadastral survey. The initial point establishes a local geographic coordinate system for the surveys that refer to that point.

An initial point is defined by the intersection of a principal meridian and a base line.

Selection of Initial Points

Figure: *Initial point marker for Utah, located at the southeast corner of Temple Square in Salt Lake City.*

A principal meridian and base line are usually established based on a preselected initial point, often some distinct geographical feature. As an example, the first established initial point in California was the Mount Diablo meridian. It was chosen because the summit of Mount

Diablo could be seen for many miles around and could be referenced for surveys.

In other cases, a meridian and a base line were chosen separately based on other geographical features.

For example, the Fifth principal meridian in the new Louisiana Purchase was established in 1815, with its southern end based on the confluence of the Arkansas and Mississippi Rivers, and extended northward from that point.

The eastern end of the base line was chosen as the confluence of the St. Francis and Mississippi Rivers and extended westward. The initial point would be established where the meridian and the base line crossed, which turned out to be in the middle of a swamp in eastern Arkansas. This point is now located in the Louisiana Purchase State Park.

Some initial points were chosen based on existing markers.

The initial point for the Gila and Salt River meridian was established based on an existing marker that had been set up in 1851 to mark a point on the Mexico–United States border before the Gadsden Purchase. Many of the initial points in the United States have been listed on the National Register of Historic Places.

Level Staff

A level staff, also called levelling rod, is a graduated wooden or aluminium rod, the use of which permits the determination of differences in elevation.

Rod Construction and Materials

Levelling rods can be one piece, but many are sectional and can be shortened for storage and transport or lengthened for use. Aluminium rods may adjust length by telescoping sections inside each other, while wooden rod sections are attached to each other with sliding connections or slip joints.

There are many types of rods, with names that identify the form of the graduations and other characteristics. Markings can be in imperial or metric units. Some rods are graduated on only one side while others are marked on both sides.

If marked on both sides, the markings can be identical or, in some cases, can have imperial units on one side and metric on the other.

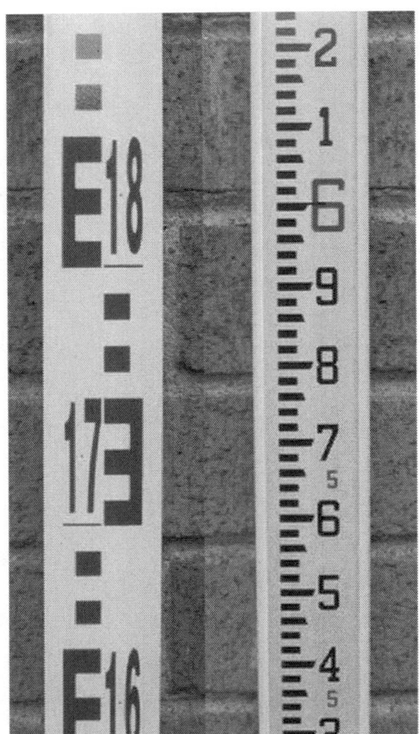

Figure: *Two sides of a modern surveyor's levelling rod. Metric graduations on the left, imperial on the right.*

Reading a Rod

In the photograph on the right, both a metric (left) and imperial (right) levelling rod are seen. This is a two-sided aluminium rod, coated white with markings in contrasting colours. The imperial side has a bright yellow background. Commonly used are metric rods.

The metric rod has major numbered graduations in metres and tenths of metres (e.g. 18 is 1.8 m – there is a tiny decimal point between the numbers). Between the major marks are either a pattern of squares and spaces in different colours or an *E* shape (or its mirror image) with horizontal components and spaces between of equal size. In both parts of the pattern, the squares, lines or spaces are precisely one centimetre high.

When viewed through an instrument's telescope, the observer can easily visually interpolate a 1 cm mark to a quarter of its height, yielding a reading with accuracy of 2.5 mm. On this side of the rod, the colours of the markings alternate between red and black with each metre of length.

The imperial graduations are in feet (large red numbers), tenths of a foot (small black numbers) and hundredths of a foot (unnumbered marks or spaces between the marks). The tenths of a foot point is indicated by the top of the long mark with the upward sloped end. The point halfway between tenths of a foot marks is indicated by the bottom of a medium length black mark with a downward sloped end. Each mark or space is approximately 3mm, yielding roughly the same accuracy as the metric rod.

Classes of Rods

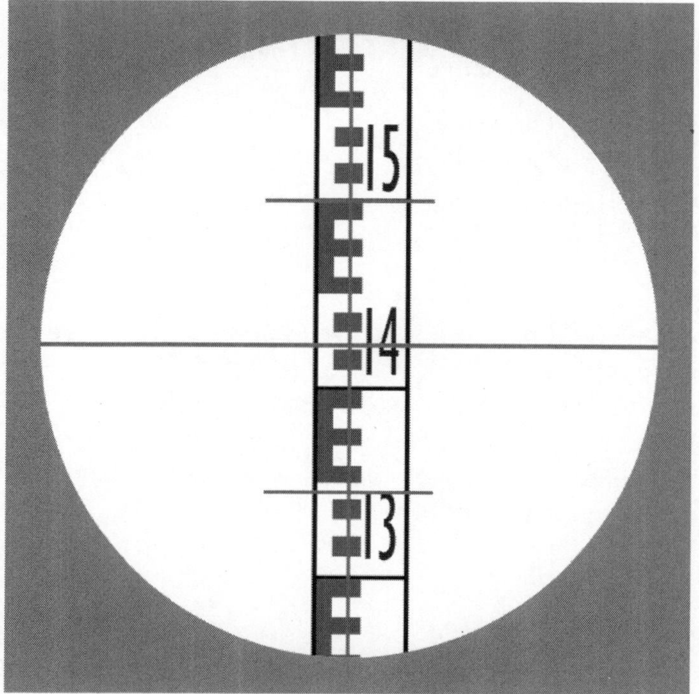

Figure: *Surveyor's view of the levelling rod with the crosshair. This indicates a reading of 1.422 m*

Rods come in two classes:

1. Self-reading rods (sometimes called speaking rods).
2. Target rods.

Self-reading rods are rods that are read by the person viewing the rod through the telescope of the instrument. The gradations are sufficiently clear to read with good accuracy. Target rods, on the other hand, are equipped with a target. The target is a round or oval plate marked in quarters in contrasting colours such as red and white in opposite quarters. A hole in the centre allows the instrument user to see the rod's scale. The target is adjusted by the rodman according to the instructions from the instrument man. When the target is set to align with the crosshairs of the instrument, the rodman records the level value. The target may have a vernier to allow fractional increments of the graduation to be read.

Topographer's Rods

Topographer's rods are special purpose rods used to ease conducting topographical surveys. The rod has the zero mark at mid-height and the graduations increase in both directions away from the mid-height.

In use, the rod is adjusted so that the zero point is level with the instrument (or the surveyor's eye if he is using a hand level for low-resolution work). When placed at any point where the level is to be read, the value seen is the height above or below the viewer's position.

An alternative topographer's rod has the graduations numbered upwards from the base.

Stadia Rods

A normal levelling rod can be used for stadia measures at shorter distances (up to about 125 m). For longer distances, special stadia rods are better suited. In order to provide good visibility at long distances, stadia rods are typically wider than levelling rods with larger markings. Since very fine gradations are not necessary for long sights, they may be left off the dedicated stadia rod.

7

Baseline Acoustic Positioning System

Long Baseline Acoustic Positioning System

A long baseline (LBL) acoustic positioning system is one of three broad classes of underwater acoustic positioning systems that are used to track underwater vehicles and divers.

Figure: *Method of the operation of a long baseline (LBL) acoustic positioning system*

The other two classes are ultra short baseline systems (USBL) and short baseline systems (SBL). LBL systems are unique in that they use networks of sea-floor mounted baseline transponders as reference points for navigation. These are generally deployed around the perimeter of a work site. The LBL technique results in very high positioning accuracy and position stability that is independent of water depth. It is generally better than 1-metre and can reach a few centimeters accuracy. LBL systems are generally employed for precision underwater survey work where the accuracy or position stability of ship-based (SBL, USBL) positioning systems does not suffice.

Operation and Performance

Figure: *A dive team (Envirotech Diving) with their AquaMap LBL acoustic underwater positioning system including three baseline transponders (B) and diver stations (A) mounted on scooters. The baseline stations are first deployed in the corners of a work site. Their relative position is then precisely measured using the AquaMap system's automatic acoustic self-survey capability. For geo-referenced operations, the baseline positions are surveyed by differential GPS or a laser positioning equipment (total station). During a dive, the diver station interrogates the baseline stations to measure the distances, which are then converted to positions.*

Long baseline systems determine the position of a vehicle or diver by acoustically measuring the distance from a vehicle or diver interrogator to three or more seafloor deployed baseline transponders.

These range measurements, which are often supplemented by depth data from pressure sensors on the devices, are then used to triangulate the position of the vehicle or diver. A diver mounted interrogator (A) sends a signal, which is received by the baseline transponders (B, C, D). The transponders reply, and the replies are received again by the diver station (A). Signal run time measurements now yield the distances A-B, A-C and A-D, which are used to compute the diver position by triangulation or position search algorithms.

The resulting positions are relative to the location of the baseline transducers. These can be readily converted to a geo-referenced coordinate system such as latitude/longitude or UTM if the geo-positions of the baseline stations are first established.

Long baseline systems get their name from the fact that the spacing of the baseline transponders is long or similar to the distance between the diver or vehicle and the transponders. That is, the baseline transponders are typically mounted in the corners of an underwater work site within which the vehicle or diver operates. This method yields an ideal geometry for positioning, in which any given error in acoustic range measurements produce only about an equivalent position error. This compares to SBL and USBL systems with shorter baselines where ranging disturbances of a given amount can result in much larger position errors.

Further, the mounting of the baseline transponders on the sea floor eliminates the need for converting between reference frames, as is the case for USBL or SBL positioning systems mounted on moving vessels. Finally, sea floor mounting makes the positioning accuracy independent of water depth. For these reasons LBL systems are generally applied to tasks where the required standard of positioning accuracy or reliability exceeds the capabilities of USBL and SBL systems.

History

The search and inspection of the lost nuclear submarine USS Thresher by the U.S. Navy oceanographic vessel USNS Mizar in 1963 is frequently credited as the origin of modern underwater acoustic navigation systems. Mizar primarily used a short baseline (SBL) system to track the bathyscaphe Trieste 1. However, its capability also included seafloor transponders, which in conjunction with early navigation satellites supported station keeping with a precision of about 300 feet, considered remarkable at the time.

Examples

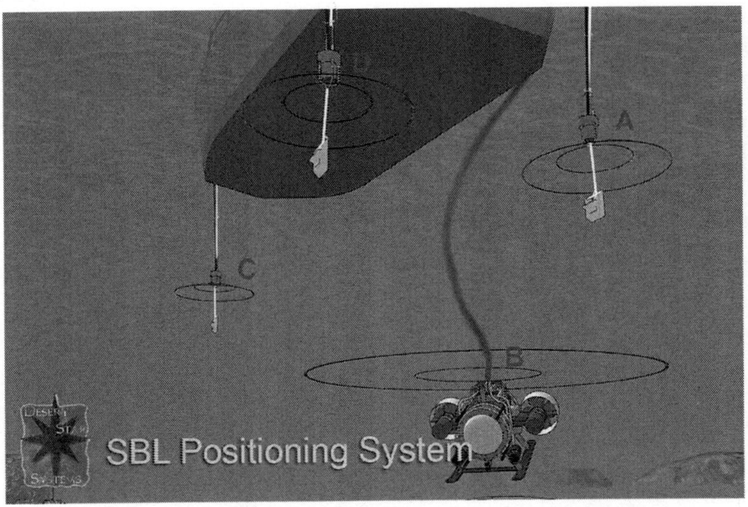

***Figure:** Precisely establishing the position of nuclear submarines prior to missile launches was an early application of long baseline acoustic positioning systems. Covert networks of sea floor transponders could survive and provide a precision navigation capability even after GPS satellites had been knocked out.*

By the mid-1960s and possibly earlier, the Soviets were developing underwater navigation systems including seafloor transponders to allow nuclear submarines to operate precisely while staying submerged. Besides navigating through canyons and other difficult underwater terrain, there was also a need to establish the position of the submarine prior to the launch of a nuclear missile (ICBM). In 1981, acoustic positioning was proposed as part of the U.S. military's MX missile system. A network of 150 covert transponder fields was envisioned. Submarines typically are guided by inertial navigation systems, but these dead reckoning systems develop position drift which must be corrected by occasional position fixes from a GPS system. If the enemy were to knock out the GPS satellites, the submarine could rely on the covert transponder network to establish its position and programme the missile's own inertial navigation system for launch.

Precise Positioning

Precise positioning is the techniques to obtain the location of an object to better than a few centimeters of accuracy.

Historically precise positioning was associated with surveying and geodesy. With the advent of Global Navigation Satellite Systems

(GNSS) such as the Global Positioning System (GPS) precise positioning has been incorporated into production processes in mining, agriculture and construction. The main application has been in machine guidance and machine automation which require high levels of precision.

Precise positioning is also increasingly used in the fields of robotics and autonomous navigation.

Short Baseline Acoustic Positioning System

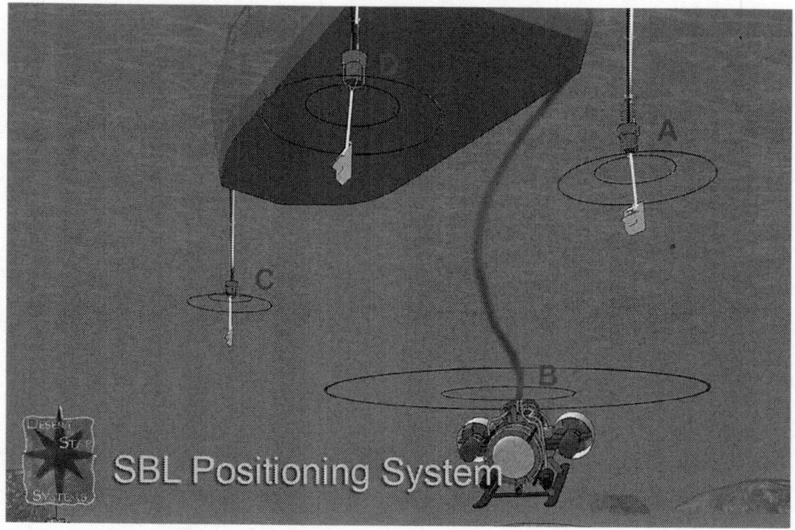

Figure: *Method of operation of a short baseline (SBL) acoustic positioning system for ROV*

A short baseline (SBL) acoustic positioning system is one of three broad classes of underwater acoustic positioning systems that are used to track underwater vehicles and divers. The other two classes are ultra short baseline systems (USBL) and long baseline systems (LBL). Like USBL systems, SBL systems do not require any seafloor mounted transponders or equipment and are thus suitable for tracking underwater targets from boats or ships that are either anchored or under way. However, unlike USBL systems, which offer a fixed accuracy, SBL positioning accuracy improves with transducer spacing. Thus, where space permits, such as when operating from larger vessels or a dock, the SBL system can achieve a precision and position robustness that is similar to that of sea floor mounted LBL systems, making the system suitable for high-accuracy survey work. When operating from a smaller vessel where transducer spacing is limited (i.e. when the baseline is short), the SBL system will exhibit reduced precision.

Operation and Performance

Short baseline systems determine the position of a tracked target such as a ROV by measuring the target's distance from three or more transducers that are, for example, lowered over the side of the surface vessel from which tracking operations take place. These range measurements, which are often supplemented by depth data from a pressure sensor, are then used to triangulate the position of the target. Baseline transducer (A) sends a signal, which is received by a transponder (B) on the tracked target. The transponder replies, and the reply is received by the three baseline transducers (A, C, D). Signal run time measurements now yield the distances B-A, B-C and B-D. The resulting target positions are always relative to the location of the baseline transducers. In cases where tracking is conducted from a moving boat but the target position must be known in earth coordinates such as latitude/longitude or UTM, the SBL positioning system is combined with a GPS receiver and an electronic compass, both mounted on the boat. These instruments determine the location and orientation of the boat, which are combined with the relative position data from the SBL system to establish the position of the tracked target in earth coordinates.

Short baseline systems get their name from the fact that the spacing of the baseline transducers (on a boat for example) is usually much less than the distance to the target, such as a robotic vehicle or diver venturing far from the boat As with any acoustic positioning system, a larger baseline yields better positioning accuracy. SBL systems use this concept to an advantage by adjusting transducer spacing for best results When operating from larger ships, from docks or from the sea ice where greater transducer spacing can be used, SBL systems can yield a positioning accuracy and robustness approaching that of sea-floor mounted LBL systems.

History

SBL systems are found employed in a variety of often specialized applications. Perhaps the first implementation of any underwater acoustic positioning system was a SBL system installed on the U.S. Navy oceanographic vessel USNS Mizar. In 1963, this system guided the bathyscape Trieste 1 to the wreck site of the American nuclear submarine USS Thresher. However, performance was still so poor that out of ten search dives by Trieste 1, visual contact was only made once with the wreckage.

The Woods Hole Oceanographic Institution is using a SHARPS SBL system to guide their JASON tethered deep ocean robotic vehicle relative to the MEDEA depressor weight and docking station associated with the vehicle. Rather than tracking both vehicles with a positioning system from the surface which would result in degraded accuracy as the pair's deployment distance, the SBL baseline transducers are mounted on MEDEA. yielding the position of JASON relative to MEDEA with good accuracy independent of the system's deployment depth. The reported accuracy is 0.09m

Example

Figure: *The SCINI ROV next to its dive hole at Heald Island, Antarctica*

An example of SBL technology is currently (since 2007) underway in Antarctica, where the Moss Landing Marine Laboratory is using a PILOT SBL system to guide the SCINI remotely operated vehicle. SCINI is a small, torpedo-shaped tethered vehicle (ROV) designed for rapid and uncomplicated deployment and exploration of remote sites around Antarctica, including Heald Island, Cape Evans and Bay of Sails. SCINI system is designed to be compact and light-weight so as to facilitate rapid deployment by helicopter, tracked vehicle and even man-hauled sleds. Once on site, its torpedo shaped body allows it to access the ocean through small (20 cm dia.) holes drilled into the sea ice. The mission's science goals however demand high accuracy in navigation, to support tasks including running 10-m video transects (straight lines), providing precise positions for still images to document the distribution and population density of benthic organisms and marking and re-visiting sites for further investigation.

The SBL navigation system consists of three small, 5 cm diameter sonar baseline transducers (A, B, C) that are linked by cable to a control box (D). A small (13.5 cm L x 4 cm D), cylinder shaped transponder is mounted on the SCINI vehicle. Accuracy is optimized by making use of the flat sea ice to place the baseline transducers well apart; approx. 35m for most SCINI deployments.

Figure reviews SCINI operations guided by the SBL system. Figure is an improvised ROV control room, in this case in a cabin hauled on top of an ice hole at Cape Armitage. From left, the displays are the ROV controls screen (A), the main camera view (B), the navigation screen (C) and the science display (D). The ROV pilot will generally watch the main camera view. He will glance at the navigation screen (C), which shows the current ROV position and track overlaid on a chart, for orientation and to guide the ROV to the location instructed by the scientist. The scientist, shown here seated on the right is provided with the science display (D), which combines the ROV imagery with position, depth and time data in real time. The scientist types written or speaks audible observations into the computer to provide a context for the data, note objects or evens of interest or designate the start or conclusion of a video transect.

A typical investigation of a site will span several dives, as tasks such as initial investigation, still image acquisition and video transects are gradually completed. A critical element in these dive series is to show prior-dive search coverage, so that a successive dive can be targeted at a previously unvisited area. This is done by producing a cumulative coverage plot of the dive site. The plot, which is updated after every dive, is displayed as a background map on the navigation screen thus providing guidance for the ongoing dive. It shows the prior ROV tracks with colour used to indicate depth. Analysis of the track data displayed here yields the quality of positioning to provide a margin of error for measurements. In this case, the typical precision has been established as 0.54m.

Survey Marker

Survey markers, also called survey marks, and sometimes geodetic marks, are objects placed to mark key survey points on the Earth's surface. They are used in geodetic and land surveying. Informally, such marks are referred to as benchmarks, although strictly speaking the term "benchmark" is reserved for marks that indicate elevation. Horizontal position markers used for triangulation are also known as triangulation stations.

Figure: *A common type of marker for a triangulation station. The triangle at its centre indicates that it is a "station mark."*

Figure: *A reference mark disk for the main station shown above. This is one of six such marks for this station.*

All sorts of different objects, ranging from the familiar brass disks to liquor bottles, clay pots, and rock cairns, have been used over the years as survey markers. Some truly monumental markers have been used to designate tripoints, or the meeting points of three or more countries. In the 19th Century, these marks were often drill holes in rock ledges, crosses or triangles chiselled in rock, or copper or brass bolts sunk into bedrock.

Today (in the United States), the most common geodetic survey marks are cast metal disks (with stamped legends on their face) set in rock ledges, sunken into the tops of concrete pillars, or affixed to the tops of pipes that have been sunk into the ground. These marks are intended to be permanent, and disturbing them is generally prohibited by federal and state law.

These marks were often placed as part of triangulation surveys, measurement efforts that moved systematically across states or regions, establishing the angles and distances between various points. Such surveys laid the basis for map-making in the United States and across the world.

Geodetic survey markers were often set in groups. For example, in triangulation surveys, the primary point identified was called the triangulation station, or the "main station". It was often marked by a "station disk", a brass disk with a triangle inscribed on its surface and an impressed mark that indicated the precise point over which a surveyor's plumb-bob should be dropped to assure a precise location over it. A triangulation station was often surrounded by several (usually three) reference marks, each of which bore an arrow that pointed back towards the main station. These reference marks made it easier for later visitors to "recover" (or re-find) the primary ("station") mark. Reference marks also made it possible to replace (or reset) a station mark that had been disturbed or destroyed.

Some old station marks were buried several feet down (to protect them from being struck by plows). Occasionally, these buried marks had surface marks set directly above them.

In the U.S., survey marks that meet certain standards for accuracy are part of a national database that is maintained by the National Geodetic Survey (NGS). Each station mark in the database has a PID (Permanent IDentifier), a unique 6-character code that can be used to call up a datasheet describing that station. The NGS has a web-based form that can be used to access any datasheet, if the station's

PID is known. Alternatively, datasheets can be called up by station name. A typical datasheet has either the precise or the estimated coordinates. Precise coordinates are called "adjusted" and result from precise surveys. Estimated coordinates are termed "scaled" and have usually been set by locating the point on a map and reading off its latitude and longitude. Scaled coordinates can be as much as several thousand feet distant from the true positions of their marks.

In the U.S., some survey markers have the latitude and longitude of the station mark, a listing of any reference marks (with their distance and bearing *from* the station mark), and a narrative (which is updated over the years) describing other reference features (e.g., buildings, roadways, trees, or fire hydrants) and the distance and/or direction of these features from the marks, and giving a history of past efforts to recover (or re-find) these marks (including any resets of the marks, or evidence of their damage or destruction).

Current best practice for stability of new survey markers is to use a punch mark stamped in the top of a metal rod driven deep into the ground, surrounded by a grease filled sleeve, and covered with a hinged cap set in concrete. Survey markers are now often used to set up a GPS receiver antenna in a known position for use in Differential GPS surveying.

Tape Correction (Surveying)

In surveying, tape correction(s) refer(s) to specific mathematical technique(s) used to apply corrections to a taping operation. Tape correction is applied to systematic or instrument errors or combination of both.

Correction Due to Incorrect Tape Length

Manufacturers of measuring tapes do not usually guarantee the exact length of tapes, and standardization is a process where a standard temperature and tension are determined at which the tape is the exact length. The nominal length of tapes can be affected by physical imperfections, stretching or wear. Constant use of tapes cause wear, tapes can become kinked and may be improperly repaired when breaks occur.

The correction due to tape length is given by:

$$C_L = M_L \pm Corr \times \frac{M_L}{N_L}$$

Where:

C_L is the corrected length of the line to be measured or laid out;

M_L is the measured length or length to be laid out;

N_L is the nominal length of the tape as specified by its mark.

Note that incorrect tape length introduces a systematic error that must be calibrated periodically.

Correction Due to Slope

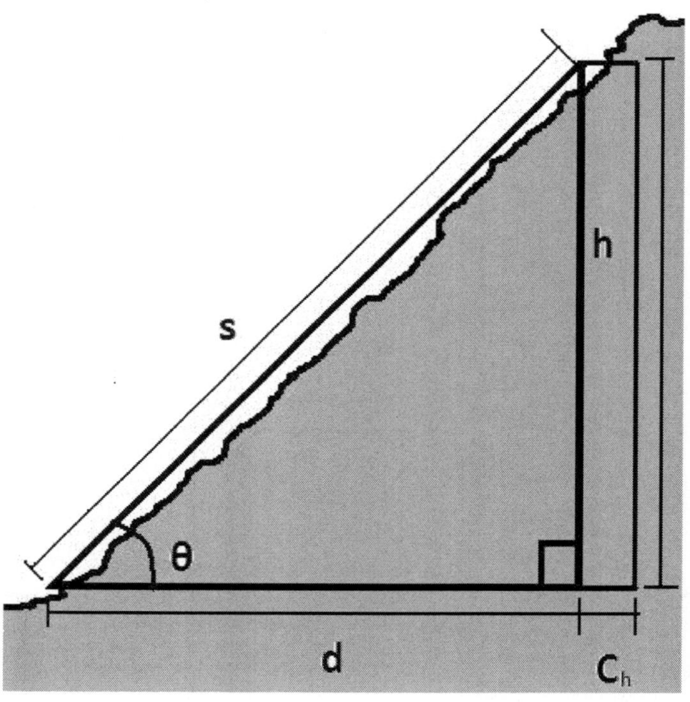

Figure: *Correction due to slope.* C_h *is the correction of height due to slope,* θ *is the angle formed by the slope line oriented from the horizontal ground,* s *is the measured slope distance between two points on the slope line,* h *is the height of the slope.*

When distances are measured along the slope, then the equivalent horizontal distance may be determined by applying a slope correction.

When applying corrections due to slope, it is necessary to determine the slope angle of the length measured. Thus,

- For gentle slopes, $m < 20\%$

$$C_h = \frac{h^2}{2s}$$

- For steep slopes, $20\% \le m \le 30\%$

$$C_h = \frac{h^2}{2s} + \frac{h^4}{8s^3}$$

- For very steep slopes, $m > 30\%$

$$C_h = s(1 - \cos\theta)$$

Where:

C_h is the correction of measured slope distance due to slope;

θ is the angle between the slope line and horizontal line;

s is the measured slope distance.

The correction C_h is subtracted from s to obtain the equivalent horizontal distance on the slope line:

$$d = s - C_h$$

Correction Due to Temperature

When measuring or laying out distances, there is always a change in temperature especially when the taping operation requires time to do so. Usually, to avoid circumstances where there is an introduced error due to temperature, tapes were standardized as a response to such factor, and a standard temperature for the tape determined.

The correction of the tape length due to change in temperature is given by:

$$C_f = C{\cdot}L(T - T_s)$$

Where:

C_f is the correction to be applied to the tape due to temperature;

T is the observed temperature or average observed temperature at the time of measurement;

T_s is the standard temperature, the temperature at which the tape was standardized;

C is the coefficient of thermal expansion of the tape;

L is the length of the tape or length of the line measured.

The correction C_f is added L to obtain the corrected distance:

$$d = L + C_f$$

Usually, for common tape measurements, the tape used is a steel tape with coefficient of thermal expansion C equal to 0.0000116 units per unit length per degree Celsius change. This means that the tape changes length by 1.16 mm per 10 m tape per 10°C change from the standard temperature of the tape.

Correction Due to Tension

Tension introduces error when the tape is pulled at a force that differs from the standard tension used at standardization. It will stretch less than its standard length when an insufficient pull is applied making the tape too short.

The tape stretches in an elastic manner (up until it reaches its elastic limit where it will deform permanently, essentially ruining the tape).

The correction due to tension is given by:

$$C_p = \frac{(P_m - P_s)L}{AE}$$

Where:

C_p is the total elongation in tape length due to pull; or the correction to be applied due to incorrect pull applied on the tape; metres;

P_m is the pull applied to the tape during measurement; kilograms;

P_s is the standard tension, it is the pull applied to the tape during standardization; kilograms;

A is the cross-sectional area of the tape; square centimeters;

E is the modulus of elasticity of the tape material; kilogram per square centimeter;

L is the measured or erroneous length of the line; metres

The correction C_p is added to L to obtain the corrected distance:

$$d = L + C_p$$

The value for A is given by:

$$A = \frac{W}{(L)(U_w)}$$

Where:

W is the total weight of the tape; kilograms;

U_w is the unit weight of the tape; kilogram per cubic centimeter.

For steel tapes, the value for U_w is given by $7.866 \times 10^{-3} kg / cm^3$.

Correction Due to Sag

A tape not supported along its length will sag and form a catenary between supports.

The correction due to sag must be calculated for each unsupported stretch separately and is given by:

$$C_s = \frac{\omega^2 L^3}{24P^2}$$

Where:

C_s is the correction applied to the tape due to sag; metres;

ω is the weight of the tape per unit length; kilogram per metres;

L is the length between two ends of the catenary; metres;

P is the tension or pull applied to the tape; kilogram.

The correction C_s is subtracted from L to obtain the corrected distance:

$$d = L - C_s$$

Note that the weight of the tape per unit length is equal to the weight of the tape divided by the length of the tape:

$$\omega = \frac{W}{L}$$

so: $W = \omega L$

Therefore, we can rewrite the formula for correction due to sag by:

$$C_s = \frac{W^2 L}{24P^2}$$

Derivation (Sag)

The general formula for a catenary formed by a tape supported only at its ends is:

$$y = \frac{P}{\omega g}\cosh\left(\frac{x\omega g}{P}\right)$$

Here g is the gravitational acceleration. The arc length between two support points at x=-k/2 and x=+k/2 is found by usual methods via integration:

$$L = \int_{-k/2}^{+k/2} \sqrt{1 + (dy/dx)^2}\, dx$$

For convenience set $a = \frac{P}{\omega g}$. The integrand is simplified as follows using hyperbolic function identities:

$$\sqrt{1+(dy/dx)^2} = \sqrt{1+\left(\frac{d}{dx}\left(a\cosh\left(\frac{x}{a}\right)\right)\right)^2} = \sqrt{1+\sinh^2\left(\frac{x}{a}\right)} = \cosh\left(\frac{x}{a}\right)$$

The tape length L is then found by integrating:

$$L = \int_{-k/2}^{+k/2} \cosh\left(\frac{x}{a}\right)dx = \left[a\sinh\left(\frac{x}{a}\right)\right]_{x=-k/2}^{x=+k/2} = (2a)\sinh\left(\frac{k}{2a}\right)$$

Now the correction for tape sag is the difference between the actual span between the supports, k, and the arc length of the tape's catenary, L. Call this correction $\delta = k - L$. The absolute value of this δ correction is C_s above, the amount you would subtract from the tape measurement to get the true span distance.

A Taylor series expansion of δin terms of the quantity L is desired to give a good first approximation to the correction. In fact the first nonvanishing term in the Taylor series is cubic in L and the next nonvanishing term is to the fifth power of L. Thus a series expansion for δ is reasonable. To this end we need to find an expression for δ that contains L but not k. We already have an expression for L in terms of k, but now need to find the inverse function (for k in terms of L):

$$\frac{L}{2a} = \sinh\left(\frac{k}{2a}\right)$$

$$\sinh^{-1}\left(\frac{L}{2a}\right) = \frac{k}{2a}$$

$$k = (2a)\sinh^{-1}\left(\frac{L}{2a}\right)$$

$$\delta = k - L = (2a)\sinh^{-1}\left(\frac{L}{2a}\right) - L$$

Evaluating delta at L=0 yields zero, so there is no zero order term in the Taylor series. The first derivative of this function with respect to L is:

$$\frac{d\delta}{dL} = \frac{1}{\sqrt{\frac{L^2}{4a^2}+1}} - 1$$

Evaluated at L=0 it vanishes and so does not contribute a Taylor series term. The second derivative of δ is:

$$\frac{d^2\delta}{dL^2} = -\frac{L}{4a^2\left(\frac{L^2}{4a^2}+1\right)^{3/2}}$$

Again when evaluated at L=0 it vanishes. When evaluated at L=0 the third derivative survives however:

$$\frac{d^3\delta}{dL^3} = -\frac{\left(8a^3 - 4\text{aL}^2\right)}{\left(4a^2 + L^2\right)^{5/2}}$$

Thus the first surviving term in the Taylor series is:

$$\delta \cong \left[\frac{d^3\delta}{dL^3}\right]_{L=0} \frac{L^3}{3!} = -\frac{1}{4a^2}\frac{L^3}{6} = \frac{-L^3}{24a^2} = \frac{-L^3\omega^2 g^2}{24P^2}$$

Notice that the variable P here is the tension on the cable, whereas above P is the mass whose gravitational force (mass times gravitational acceleration) equals the tension on the cable. The only conversion necessary then is to take P/g here and equate it to P above. Also this formula is the tape sag correction to be added to the measured distance, so the negative sign in front can be removed and the tape sag correction can be made instead by subtracting the absolute value as is done in the preceding section.

Tripod (Surveying)

A surveyor's tripod is a device used to support any one of a number of surveying instruments, such as theodolites, total stations, levels or transits.

History

Older surveying tripods had slightly different features compared to modern ones. For example, on some older tripods, the instrument had its own footplate and did not need to move laterally relative to the tripod head. For this reason, the head of the tripod was not a flat footplate but was simply a large diameter fitting. Threads on the outside of the head engaged threads on the instrument's footplate. No other mounting screw was used.

Fixed length legs were also seen on older instruments. Instrument height was adjusted by changing the angle of the legs. Widely spaced tripod feet resulted in a lower instrument while closely spaced legs raised the instrument. This was considerably less convenient than having variable length legs.

Materials for older tripods were predominantly wood and brass, with some steel for high wear items like the feet or foot points.

Usage

Figure: *This shows the head of a surveyor's tripod with the mounting screw in the opening.*

The tripod is placed in the location where it is needed. The surveyor will press down on the legs' platforms to securely anchor the legs in soil or to force the feet to a low position on uneven, pock-marked pavement. Leg lengths are adjusted to bring the tripod head to a convenient height and make it roughly level.

Once the tripod is positioned and secure, the instrument is placed on the head. The mounting screw is pushed up under the instrument to engage the instrument's base and screwed tight when the instrument is in the correct position.

The flat surface of the tripod head is called the foot plate and is used to support the adjustable feet of the instrument.

Positioning the tripod and instrument precisely over an indicated mark on the ground or benchmark requires intricate techniques.

Figure: *This shows a surveyor's tripod's foot. The platform is used to push the spike into the ground. Above the foot is the height adjustment.*

Construction

Many modern tripods are constructed of aluminium, though wood is still used for legs. The feet are either aluminium tipped with a steel point or steel. The mounting screw is often brass or brass and plastic. The mounting screw is hollow to allow the optical plumb to be viewed through the screw. The top is typically threaded with a 5/8" x 11 tpi screw thread. The mounting screw is held to the underside of the tripod head by a movable arm. This permits the screw to be moved anywhere within the head's opening. The legs are attached to the head with adjustable screws that are usually kept tight enough to allow the legs to be moved with a bit of resistance. The legs are two part, with the lower part capable of telescoping to adjust the length of the leg to suit the terrain. Aluminium or steel slip joints with a tightening screw are at the bottom of the upper leg to hold the bottom part in place and fix the length. A shoulder strap is often affixed to the tripod to allow for ease of carrying the equipment over areas to be surveyed.

Circular Curves in Engineering Survey

Advertisements

In the geometric design of motorways, railways, pipelines, etc., the design and setting out of curves is an important aspect of the engineer's work. The initial design is usually based on a series of straight sections whose positions are defined largely by the topography of the area. The intersections of pairs of straights are then connected by horizontal curves.

In the vertical design, intersecting gradients are connected by curves in the vertical plane.

Circular curves are used to join intersecting straight lines (or tangents). Circular curves are assumed to be concave. Horizontal circular curves are used to transition the change in alignment at angle points in the tangent (straight) portions of alignments. An angle point is called a point of intersection or PI station; and, the change in alignment is defined by a deflection angle, Δ.

Types of Circular Curves Are:

1. Simple Curve
2. Compound Curves
3. Broken Back Curves
4. Reverse Curves

Radius of a Circular Curve

The Radius is the distance from the centre of the curve to any point on the circular curve.

Direction of a Circular Curve

The Direction of a Circular Curve is defined as the direction the curve tends, as stationing along the curve increases. Can be expressed as: Left, Right, North, East, South, West, free text

Central Angle of a Circular Curve

The Central Angle of a Circular Curve is the angle at the centre of radius of a circular arc included between the radii, passing through the beginning and ending of the arc.

Long Chord Length

The Long Chord Length is the straight line distance connecting the beginning of the curve and the end of the curve.

Degree of Curvature

The Degree of Curve is defined as the angle subtended by an arc whose length is 100 ft. A Radian is the angle subtended by an arc whose length equals the length of the Radius, or

57° 17' 44.8" , or 57.295779513°.

i. Curvature can be expressed in two ways, By:

1. Stating the length of the chord of the curve
2. Stating the radius of curvature

Laying out Circular Curves

1. Select tangents, and general curves making sure you meet minimum radius criteria
2. Select specific curve radii/spiral and calculate important points using formula or table (those needed for design, plans, and lab requirements)
3. Station alignment (as curves are encountered)
4. Determine super and runoff for curves and put in table
5. Add information to plans

Sight Distance of a Circular Curve

Sight line is a chord of the circular curve. Sight Distance is curve length measured along centerline of inside lane. Sight distance can be the controlling aspect of horizontal curve design where obstructions are present near the inside of the curve. To determine the actual sight distance that you have provided, you need to consider that the driver can only see the portion of the roadway ahead that is not hidden by the obstruction. In addition, at the instant the driver is in a position to see a hazard in the roadway ahead, there should be a length of roadway between the vehicle and the hazard that is greater than or equal to the stopping sight distance

Curves should be designed with their radius greater than R_{min}. If R_{min} cannot provided enough lateral clearance to an obstruction.

Route Survey

Definition: A Route Survey is defined as being the required service and product that adequately locates the planned path of a linear project or right of way which crosses a prescribed area of real estate, extending from at least one known point and turning or terminating at another known point. Adequate location shall mean substantial compliance with the conditions and tolerances expressed in this standard. A route survey which defines new or proposed boundaries shall be conducted as a boundary survey and must adhere to the rules and regulations of the Texas Board of Professional Land Surveying (TBPLS).

Purposes: A Route Survey is usually required for the planning of a right of way, the acquisition of fee or easement property and for eventual construction layout work. The location of the facilities within the right of way are often held in respect to the centre line or a right of way line. A Route Survey is made on the ground to provide for the location of right of way lines, a centerline, or reference lines in relation to property lines and terrain features.

Route Surveys shall include but are not limited to the proper location, monumentation, description or platting of the following routes:

- Roadways, highways and railroads.
- Transmission lines for communications, fuel, chemical, water and electrical needs.
- Canals, waterways, drainage ditches and sewers.
- View easements, air space easements, ingress and egress easements such as approach routes.

Mine Surveying

Mine surveying is a branch of mining science and technology. It includes all measurements, calculations and mapping which serve the purpose of ascertaining and documenting information at all stages from prospecting to exploitation and utilizing mineral deposits both by surface and underground working.

The following are the principal activities of Mine surveying:

- The interpretation of the geology of mineral deposits in relation to the economic exploitation thereof
- The investigation and negotiation of mineral mining rights
- Making and recording, and calculations of mine surveying measurements
- Mining cartography
- Investigation and prediction of effects of mine working on the surface and underground strata
- Mine planning in the context of local environment and subsequent rehabilitation

The activities involve:

- The location, structure, configuration, dimensions and characteristics of the mineral deposits and of the adjoining rocks and overlying strata. The assessment of mineral reserves and the economics of their exploitation
- The acquisitation, sale, lease and management of mineral properties
- Providing the basis of the planning, direction and control of mine workings to ensure economical and safe mining operations
- The study of rock and ground movements caused by mining operations, their prediction, and the precautions and remedial treatment of subsidence damage
- Assisting in planning and rehabilitation of land affected by mineral operations and collaborating with local government planning authorities

8

Astronomical Survey

An astronomical survey is a general map or image of a region of the sky which lacks a specific observational target. Alternatively, an astronomical survey may comprise a set of many images or spectra of objects which share a common type or feature. Surveys are often restricted to one band of the electromagnetic spectrum due to instrumental limitations, although multiwavelength surveys can be made by using multiple detectors, each sensitive to a different bandwidth. Surveys have generally been performed as part of the production of an astronomical catalogue.

Scientific Value

Sky surveys, unlike targeted observation of a specific object, allow astronomers to catalogue celestial objects and perform statistical analyses on them without making prohibitively lengthy observations. In some cases, an astronomer interested in a particular object will find that survey images are sufficient to entirely obviate the need for telescope time.

Surveys also help astronomers obtain observation time on larger, more powerful telescopes. If the astronomer can show a telescope scheduling committee that previous observations support his hypothesis, he is more likely to be given a chance to make more detailed observations.

The wide scope of surveys makes them ideal for astronomers searching for moving foreground objects such as asteroids and comets. An astronomer can compare existing survey images to current observations to locate targets which are in motion; this task can even be performed automatically using image analysis software. Similarly, images of the same object taken by different surveys can be compared to detect transient events such as variable stars.

List of Sky Surveys

- Optical
  - Pan-Andromeda Archaeological Survey
  - National Geographic Society – Palomar Observatory Sky Survey (NGS–POSS) - survey of the northern sky on photographic plates, 1948–1958
  - CfA Redshift Survey A programme from Harvard-Smithonian Centre for Astrophysics. It began in 1977 to 1982 then from 1985 to 1995.
  - Digitized Sky Survey - optical all-sky survey created from digitized photographic plates, 1994
  - Sloan Digital Sky Survey - an optical and spectroscopic survey, 2000-2006 (first pass)
  - Photopic Sky Survey - a survey with 37,440 individual exposures, 2010-2011.
  - Palomar Distant Solar System Survey (PDSSS)
  - WiggleZ Dark Energy Survey(2006-2011)used the Australian Astronomical Observatory
  - Dark Energy Survey(DES) Is a survey about one-tenth of the sky to find clues to the characteristics of dark energy.-
- Infrared
  - Infrared Astronomical Satellite did an all sky survey at 12, 25, 60, and 100 μm, 1983
  - The 2-micron All-Sky Survey (2MASS), a ground based all sky survey at J, H, and Ks bands (1.25, 1.65, and 2.17 μm) 1997-2001
  - Akari (Astro-F) a Japanese mid and far infrared all-sky survey satellite, 2006–2008
  - Wide-field Infrared Survey Explorer was launched in December 2009 to begin a survey of 99% of the sky at wavelengths of 3.3, 4.7, 12, and 23 μm. The telescope is over a thousand times as sensitive as previous infrared surveys. The initial survey, consisting of each sky position imaged at least eight times, was completed by July 2010.

  - SCUBA-2 All Sky Survey
  - VISTA - Several Surveys (VVV, VIKING, VHS, etc.)
- Radio
  - HIPASS - Radio survey, the first blind HI survey to cover the entire southern sky. 1997-2002
  - Ohio Sky Survey - Over 19,000 radio sources at 1415 MHz. 1965-1973.
  - NVSS - Survey at 1.4 GHz mapping the sky north of -40 deg
  - FIRST - Survey to look for faint radio sources at twenty cms.
  - PALFA Survey - On-going 1.4 GHz survey for radio pulsars using the Arecibo Observatory.
  - GALEX Arecibo SDSS SurveyGASS designed to measure the neutral hydrogen content of a representative sample of ~1000 massive, galaxies
- Gamma-ray
  - Fermi Gamma-ray Space Telescope, formerly referred to as the "Gamma-ray Large Area Space Telescope (GLAST)." 2008–present; the goal for the telescope's lifetime is 10 years.
- Multi-wavelength surveys
  - GAMA - the Galaxy And Mass Assembly survey combines data from a number of ground- and space-based observatories together with a large redshift survey, performed at the Anglo-Australian Telescope. The resulting dataset aims to be a comprehensive resource for studying the physics of the galaxy population and underlying mass structures in the recent universe.
  - GOODS - The Great Observatories Origins Deep Survey.
  - COSMOS - The Cosmic Evolution Survey
  - (The latter two surveys are joining together observations obtained from space with the Hubble Space Telescope, the Spitzer Space Telescope, the Chandra X-ray Observatory and the XMM-Newton satellite, with a large set of observations obtained with ground-based telescopes).
  - Atlas 3d Survey - sample of 260 galaxies for the Astrophysics project

- Planned
  - Pan-STARRS - a proposed 4-telescope large-field survey system to look for transient and variable sources
  - Large Synaptic Survey Telescope - a proposed very large telescope designed to repeatedly survey the whole sky that is visible from its location

Surveys of the Magellanic Clouds

- The Magellanic Clouds Photometric Survey - UBVI (optical)
- Deep Near Infrared Survey (DENIS) - near-IR
- Surveying the Agents of a Galaxy's Evolution - a Spitzer Space Telescope legacy observation programme of the LMC

Traverse Computations

Traverse

A *traverse* is a series of connected lines whose lengths and adjacent angular relationships are measured. Traverses are used to determine horizontal positions, relative locations, and areas. A traditional traverse is two-dimensional in a horizontal plane.

As Total Station Instrument technology evolved, more and more traverse data was collected in all three dimensions, incorporating elevation. This topic will address only two-dimensional horizontal traverses although many of the concepts are easily adopted to three-dimensions.

Configuration

Configuration is the physical network design of a traverse; it can be either a *loop* or a *link*.

Type

Type is defined mathematically. A closed traverse has sufficient measurements to enable one or more math checks which facilitate tracking down errors. An open traverse has just enough measurements to compute point positions, but not enough to check mathematical conditions.

For example, the loop traverse is mathematically closed because all four distances and angles have been measured. Each traverse point's relative position is fixed with respect to the others. One of the traverse's mathematical conditions which can be checked is that the angles should sum exactly to 360°.

The loop traverse has only its distances measured, no angles. It is open because its polygonal configuration isn't fixed by distances alone. The traverse sides have the same lengths as those yet its polygonal configuration is different. Individual points are free to move with respect to the others since there are insufficient measurements to fix them. Should any distance be in error, there is no way to track it down. Both these examples demonstrate that a loop traverse may be closed or open - it depends on the number of measurement made. The same is true for a link traverse. The conditions which must be met for loop and link traverses to be closed.

Whenever possible, a closed traverse should be used. There are exceptions, but open traverses are generally limited to situations where lower accuracy is acceptable.

Computation Steps

Traverse computations is the process of taking field measurement through a series of mathematical calculations to determine final traverse size and configuration. These calculations include error compensation as well as reformation to determine quantities not directly measured.

Traditional traverse computation steps are:

1. Balance (adjust) angles
2. Determine line directions
3. Compute latitudes and departures
4. Adjust the traverse misclosure
5. Determine adjusted line lengths and directions
6. Compute coordinates
7. Compute area.

The order of some steps can be changed. For example, steps 1 and 2 would be reversed for closed link traverses with directions at both ends. Balancing angles would normally not be done If a least squares adjustment is used at step 4.

The complete series of computations can only be performed on closed traverses. That's because some of the steps require adjustment of errors and, as discussed before, errors can't be identified in an open traverse.

Closed loop traverse examples will be used to demonstrate the computation steps in sequence. Once the complete process is covered, we will examine how to perform the same operations, except for area computations, on closed link traverses.

A traditional approach does not need coordinates, even for area determination. Most software, however does. Whether done manually or with software, the process starts with initial point coordinates based on the raw field measurements; at each successive computational step the coordinates are updated reflecting the particular operation. We will go through a traverse adjustment by coordinates example of using one of the examples for the traditional approach for comparative purposes.

Line Directions

Meridian

A meridian is a north-south reference line. It is used as a basis for a direction, which describes a line's orientation.

Common meridians include, but are not limited to:

- True
- Magnetic
- Astronomic
- Grid
- Geodetic
- Assumed

  Depending on the system, meridians may converge or may stay parallel.

For small project areas such as a single lot survey, meridians can assumed to be parallel, regardless of system, without introducing significant error.

Most meridian systems (even Assumed) are constant over time. However, magnetic meridians change location and their change is not consistent. Magnetic meridians are important as the first instrument extensively used for property surveys was the compass. Although True directions were reported in the early Public Land Surveys, they were first measured by compass then converted using solar observation.

A surveyor will select an appropriate meridian for the project at hand and orient all survey lines to it. Sometimes directions based on one meridian must be converted to a different meridian in order to maintain consistency across surveys. We'll look at that, along with magnetic direction conversions, in *Meridian Conversions* later.

Direction

A *direction* is angle from a meridian to a line. It is similar to a horizontal angle in a traverse except the backsight is always along the meridian.

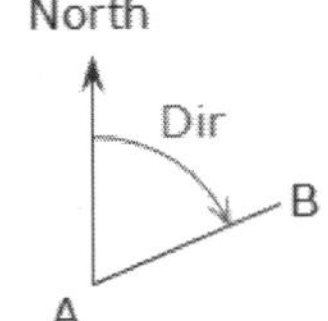

Figure:- *Direction*

There are two different ways to express line direction: bearing and azimuth.

Bearing

A bearing is an angle from the North or South end of the meridian turned to the East or West. A bearing has three parts:

- *Prefix* - N or S indicating which end of the meridian is turned from.
- *Angle*
- *Suffix* - E or W indicating turning direction from the meridian to the line.

N 66°40' E - from the North end of the meridian, turn 66°40' to the East.

Example

Bearing AB = N 66°40' E

Bearing AC = S 55°32' E

Bearing AD = S 44°21' W

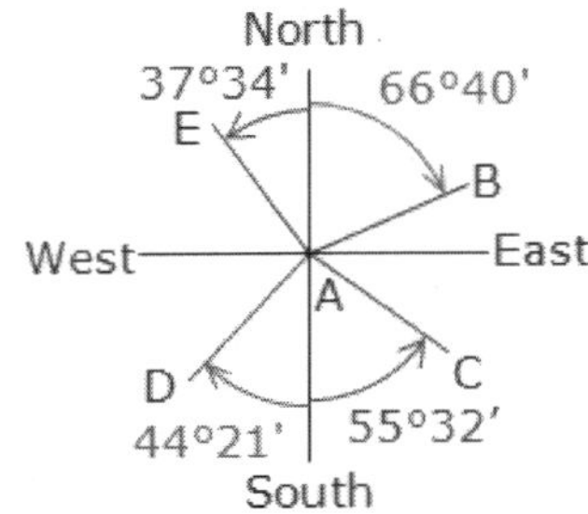

Figure:- *Bearings*

A bearing falls in one of four quadrants so the angle does not exceed 90°. The angle is to the right (clockwise) in the NE and SW quadrants, to the left (counter-clockwise) in the SE and NW quadrants. A due North direction can be expressed as either N 00°00' E or N 00°00' W; due East as N 90°00' E or S 90°00' E; similarly for dues South and West.

A back-bearing is reverse of a bearing, that is, Bearing BA is the back-bearing of Bearing AB. Because the meridians are parallel at both ends of the line, the bearing angle is the same but quadrant is reverse. This is true only when meridians are parallel. Where meridians converge, the forward and back bearing angles will differ by the total convergence. More on this later.

Bearing AB = N 66°40' E

Bearing BA = S 66°40' W

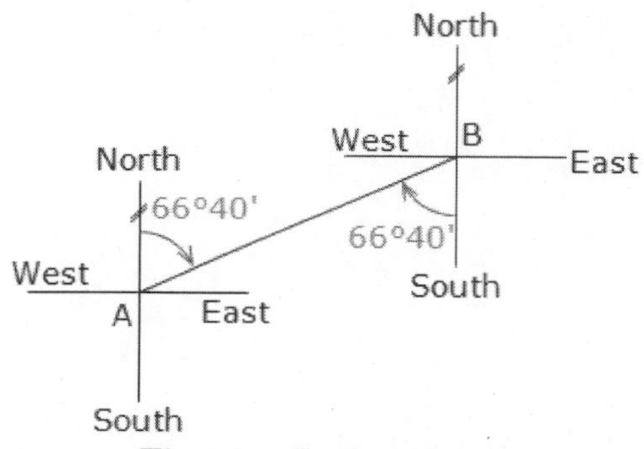

Figure:- *Back bearing*

Azimuth

An azimuth is an angle to the right (clockwise) from the meridian to the line. In most cases the azimuth is turned from the north meridian end; earlier control surveys used the south end. An azimuth varies from 0° to 360°.

Example

Azimuth AB = 66°40'

Azimuth AC = 124°28'

Azimuth AD = 224°21'

Azimuth AE = 322°26'

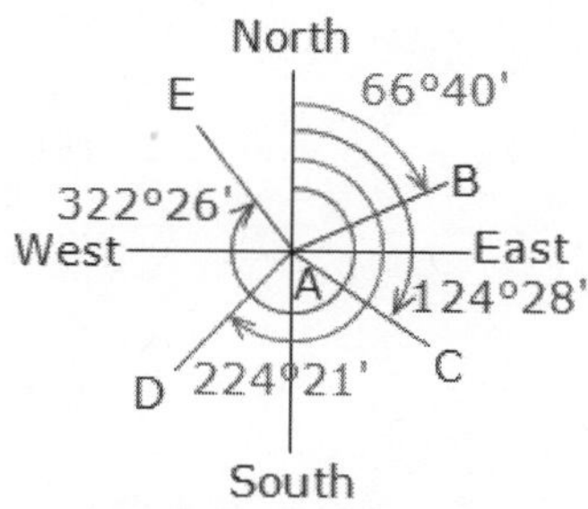

Figure:- *Azimuths*

A back-azimuth is reverse of a azimuth: Azimuth CA is the back-azimuth of Azimuth AC. Because the meridians are parallel at both ends of the line, the back-azimuth and forward azimuth differ by 180°. As with bearings, this is true only when meridians are parallel. Where meridians converge, the forward and back azimuths will differ by (180° ± total convergence). More on this later.

Example

Azimuth AC = 124°28'

Azimuth CA = 124°28' + 180°00' = 304°28'

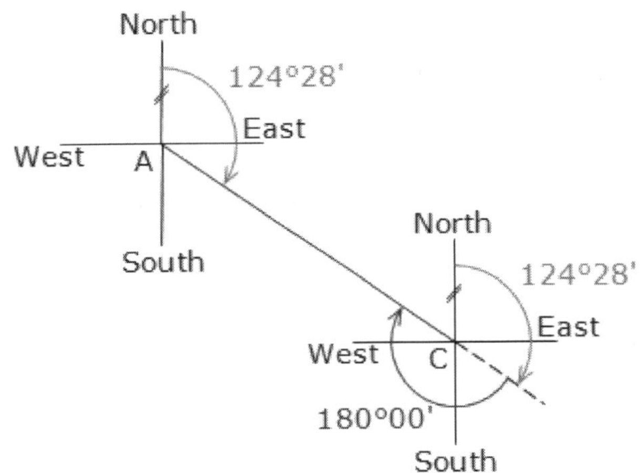

Figure:- Back azimuth

Converting Between Bearings and Azimuths

Since Bearings and Azimuths are both referenced to a meridian it is simple to convert one to the other.

To convert from bearings to azimuths:

Table:- Direction quadrants		
Quadrant	***From Bearing***	***To Azimuth***
NE	N β E	β
SE	S β E	180° - β
SW	S β W	180° + β
NW	N β W	360° - β

Example

Azimuth AB = 66°40'

Azimuth AC = 180°00' - 55°32' = 124°28'

Azimuth AD = 180°00' + 44°21' = 224°41'

Azimuth AE = 360°00' - 37°34' = 322°26'

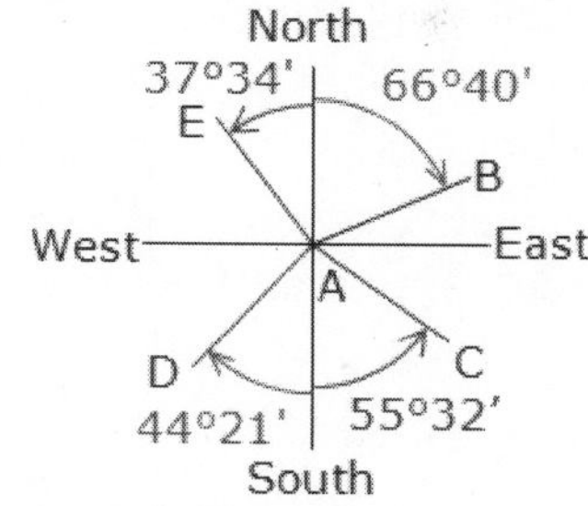

Figure:- *Bearing to azimuth*

To convert from an azimuth, α, to a bearing:

Table:- *Azimuth to bearing*

Quadrant	***To Bearing***
NE	N α E
SE	S (180° - α) E
SW	S (α - 180°) W
NW	N (360° - α) W

Example

Bearing AB = N 64°40' E

Bearing AC = S (180°00'-124°28') E = S 55°32' E

Bearing AD = S (224°21' - 180°00') W = S 44°21' W

Bearing AE = N (360°00' - 322°26') W = N 37°34' W

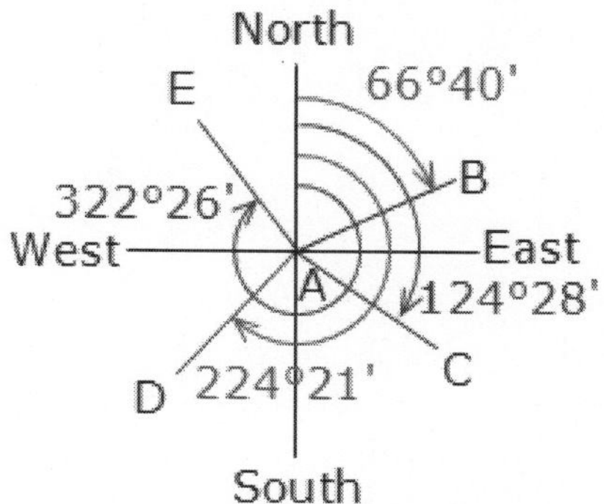

Figure:- *Azimuth to bearing*

Rather then memorize tables, drawing a sketch will help determine correct conversion logic to use.

Alidade

An alidade (archaic forms include alhidade, alhidad, alidad) or a turning board is a device that allows one to sight a distant object and use the line of sight to perform a task.

This task can be, for example, to draw a line on a plane table in the direction of the object or to measure the angle to the object from some reference point. Angles measured can be horizontal, vertical or in any chosen plane.

The alidade's primary use is for creating maps in the horizontal plane.

The alidade was originally a part of many types of scientific and astronomical instrument. At one time, some alidades, particularly those used on graduated circles as on astrolabes, were also called *diopters*. With modern technology, the name is applied to complete instruments such as the plane table alidade.

Origins

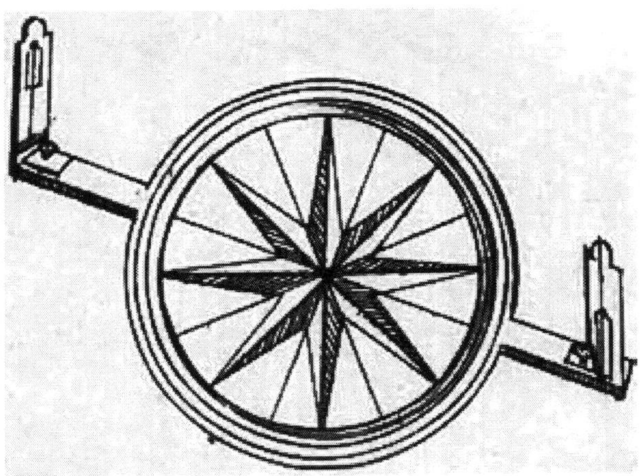

Figure: *An example of an alidade on a circumferentor. Taken from the Table of Surveying, Cyclopaedia, Volume 2, 1728*

The word in Arabic, signifies the same device. In Greek and Latin, it is respectively called "dioptra", and *linea fiduciae*, "fiducial line". The earliest alidades consisted of a bar, rod or similar component with a vane on each end. Each vane (also called a pinnule or pinule) has a hole, slot or other indicator through which one can view a distant object. There may also be a pointer or pointers on the alidade to indicate a position on a scale. Alidades have been made of wood, ivory, brass and other materials.

Examples of Old Alidade Types

The figure on the left displays drawings that attempt to show the general forms of various alidades that can be found on many antique instruments. Real alidades of these types could be much more decorative, revealing the maker's artistic talents as well as his technical skills. In the terminology of the time, the edge of an alidade at which one reads a scale or draws a line is called a *fiducial edge*.

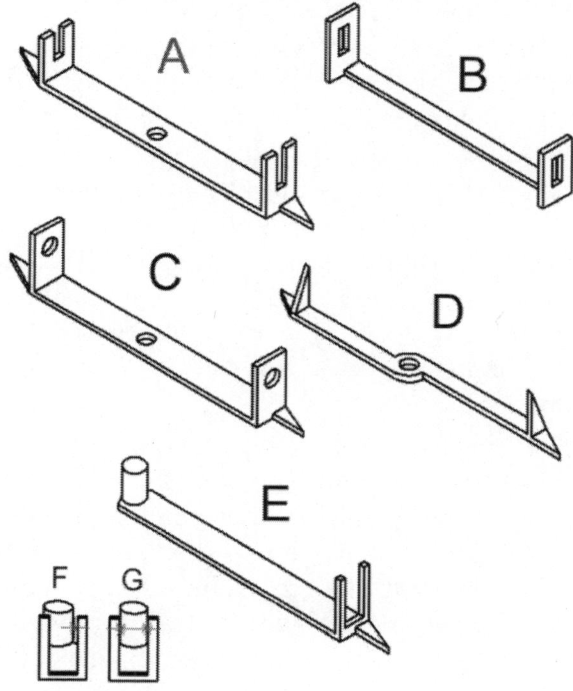

Figure: *Several examples of alidade types.*

Alidade B in the diagram shows a straight, flat bar with a vane at either end. No pointers are used. The vanes are not centred on the bar but offset so that the sight line coincides with the edge of the bar.

The vanes have a rectangular hole in each with a fine wire held vertically in the opening. To use the alidade, the user sights an object and lines it up with the wires in each vane. This type of alidade could be found on a plane table, graphometer or similar instrument.

Alidades A and C are similar to B but have a slit or circular hole without a wire. In the diagram, the openings are exaggerated in size to show the shape; they would be smaller in a real alidade, perhaps 2 mm or so in width. One can look through the openings and line the openings up with the object of interest in the distance. With a small opening, the error in sighting the object is small. However, if using a small hole to observe a dim object such as a star, the image is difficult to see. This form is shown in the diagram as having pointers. These can be used to read off an angle on a scale that is engraved around the outer edge (or limb) of the instrument. Alidades of this form are found on astrolabes, mariner's astrolabes and similar instruments.

Alidade D has vanes without any openings. In this case, the object is viewed and the alidade is rotated until the two opposite vanes simultaneously eclipse the object. With skill, this sort of alidade can yield very precise measures. In this example, pointers are shown.

Alidade E is a representation of a very interesting design by Johannes Hevelius. Hevelius was following in Tycho Brahe's footsteps and cataloguing star positions with high accuracy. He did have access to the telescopic sights that were being used by astronomers in other countries, however, he chose to use naked-eye observations for his positional instruments. Due to the design of his instruments and the alidades, as well as his diligent practices, he was able to yield very precise measures.

Hevelius' design featured a pivot point with a vertical cylinder and a vane at the observer's end. The vane had two narrow slits that were spaced precisely the same distance apart as the diameter of the cylinder (in the diagram, the portion of the vane between the slits is removed for clarity; the left and right edges of the opening represent the slits). If the observer could sight a star on only one side of the cylinder, as seen in F, the alignment was off. By carefully moving the vane so that the star could just barely be seen on either side of the cylinder (G), the alidade was aligned with the position of the star. This could not be used with a closely located object. A star, being so far away as to exhibit no parallax to the naked-eye, would be observable as a point source simultaneously on both sides.

Modern Alidade Types

- The alidade is the part of a theodolite that rotates around the vertical axis, and that bears the horizontal axis around which the telescope (or visor, in early telescope-less instruments) turns up or down.
- In a sextant the alidade is the turnable arm carrying a mirror and an index to a graduated circle in a vertical plane. Today it is more commonly called an *index arm*.

- Alidade tables have also long been used in fire towers for sighting the bearing to a forest fire. A topographic map of the local area, with a suitable scale, is oriented, centred and permanently mounted on a leveled circular table surrounded by an arc calibrated to true north of the map and graduated in degrees (and fractions) of arc. Two vertical sight apertures are arranged opposite each other and can be rotated along the graduated arc of the horizontal table. To determine a bearing to a suspected fire, the user looks through the two sights and adjusts them until they are aligned with the source of the smoke (or an observed lightning strike to be monitored for smoke).

Circumferentor

Figure: *Circumferentor with gunter's chain at Campus Martius Museum in Marietta, Ohio*

A circumferentor, or surveyor's compass, is an instrument used in surveying to measure horizontal angles, now superseded by the theodolite. It consists of a brass circle and an index, all of one piece. On the circle is a card, or compass, divided into 360 degrees; the meridian line of which is in the middle of the breadth of the index. On the circumference of the circle is a brass ring, which, with another ring fitted with glass, make a kind of box for the needle, which is suspended on a rivet in the centre of the circle. On each extreme of the index is a sight. The whole apparatus is mounted on a staff, with a ball-and-socket joint for easy rotation.

Circumferentors were made throughout Europe, including England, France, Italy, and Holland. By the early 19th century, Europeans preferred theodolites to circumferentors. However, in America, and other wooded or uncleared areas, the circumferentor was still in common use.

Usage

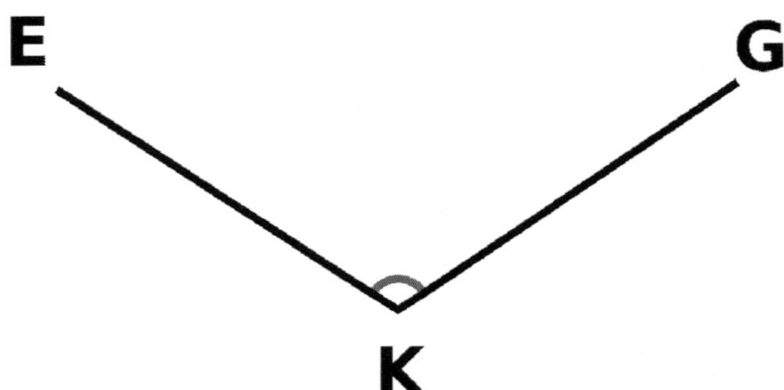

Figure: *Angle EKG*

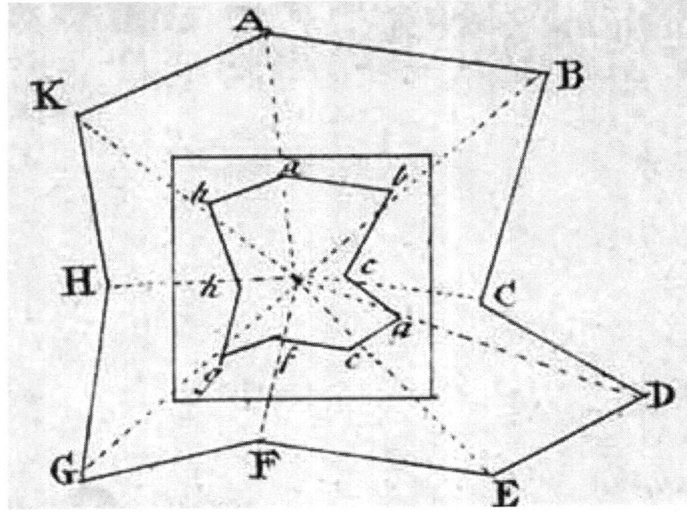

Figure: *Region ABCDEFGHK*

Measuring Angles

To measure an angle with a circumferentor, such as angle EKG, place the instrument at K, with the fleur-de-lis in the card towards you. Then direct the sights, until through them you see E; and note the degree pointed at by the south end of the needle, such as 296°. Then, turn the instrument around, with the fleur-de-lis still towards you, and direct the sights to G; note the degree at which the south end of the needle point, such as 182°.

Finally, subtract the lesser number, 182, from the greater, 296°; the remainder, 114°, is the number of degrees in the angle EKG.

If the remainder is more than 180 degrees, it must be subtracted from 360 degrees.

Surveying a Region

To take the plot of a field, forest, park, etc., with a circumferentor, consider region ABCDEFGHK, an area to be surveyed.

1. Placing the instrument at A, the fleur-de-lis towards you, direct the sights to B; where suppose the south end of the needle cuts 191°; and the ditch, wall, or hedge, measuring with a Gunter's chain, contains 10 chains, 75 links.
2. Placing the instrument at B, direct the sights as before to C; the south end of the needle, *e.g.* will cut 279°; and the line BC contains 6 chains and 83 links.

Then move the instrument to C; turn the sights to measure D, and measure CD as before. In the same manner, proceed to D, E, F, G, H, and lastly to K; still noting the degrees of every bearing, or angle, and the distances of every side. This will result in a table of the following form:

Station	*Degrees*	*Min.*	*Chains*	*Link*
A	191	00	10	75
B	297	00	6	83
C	216	30	7	82
etc.				

From this table, the field is to be plotted, or protracted.

Alternative Plotting method: An alternative way to plot the area is to use several angles and only a few measurements and calculate their positions.

This could be done by starting at the centre point in which is not labelled, but which will be referred to as "Centre." Assume each point can be seen from each other point. From the "centre" point, sight and record the angle to each point using the sights as described above. Then move to, and measure the distance to, one of the other points referenced, such as point B. At point B, measure the angles to all the other points.

Then, move to an additional point such as point F. Again, measure the distance from the centre to the point chosen (F).

At that point, measure and record the angles to each of the other points as was done at point B. Chose a scale (a ratio between the size of the area to be plotted and the size of the paper on which you will draw the plot) that will allow the plot to fit on your paper and plot the angles and distances.

The advantage of this method over the first one above is that there are fewer distance measurements and any errors in angles or distances will not be cumulative; that is, if you use the first survey method, any angle that is slightly off will distort the remainder of the plot. The second method can also be used when it is not possible to measure some of the distances, for example, if there is a water barrier between two of the points.

Also, if there are any inaccuracies in the measurements, they will be revealed in the plot because the points plotted from various angles will not coincide.

Additional considerations include the number of times the circumferentor must be set up and aligned. With the first method, the instrument must be set up at each point with a compass. With the second method, the initial set up is at "centre."

After that, for example at point B, the instrument can be set up by aligning the sight with the reciprocal of the angle between "centre" and B. Thus, any local change in the magnetic field that would affect the compass would be nullified.

Surveyor's Double Prism

A double prism is a device to measure right angles, consisting of two five sided prisms stacked on top of each other and a plumb-bob. it is used to stake out right angles, for example on a construction site.

Dumpy Level

A dumpy level, builder's auto level, levelling instrument, or automatic level is an optical instrument used to establish or check points in the same horizontal plane. It is used in surveying and building to transfer, measure, or set horizontal levels.

Invention

In 1832, Englishcivil engineer William Gravatt, who had worked with Marc Isambard Brunel and his son Isambard on the Thames Tunnel, was commissioned by Mr. H.R. Palmer to examine a scheme for the South Eastern Railway's route from London to Dover. Forced to use the then conventional Y level, during the works Gravatt devised the more transportable and easier to use dumpy level.

Operation

The level instrument is set up on a tripod and, depending on the type, either roughly or accurately set to a leveled condition using footscrews (levelling screws).

The operator looks through the eyepiece of the telescope while an assistant holds a tape measure or graduated staff vertical at the point under measurement. The instrument and staff are used to gather and/or transfer elevations (levels) during site surveys or building construction. Measurement generally starts from a benchmark with known height determined by a previous survey, or an arbitrary point with an assumed height.

Variants

The term dumpy level endures despite the evolution in design. A dumpy level is an older-style instrument that requires skilled use to set accurately. The instrument requires to be set level in each quadrant, to ensure it is accurate through a full 360° traverse. Some dumpy levels will have a bubble level ensuring an accurate level.

A variation on the dumpy and one that was often used by surveyors, where greater accuracy and error checking was required, is a tilting level. This instrument allows the telescope to be effectively flipped through 180°, without rotating the head. The telescope is hinged to one side of the instrument's axis; flipping it involves lifting to the other side of the central axis (thereby inverting the telescope). This action effectively cancels out any errors introduced by poor setup procedure or errors in the instrument's adjustment. As an example, the identical effect can be had with a standard builder's level by rotating it through 180° and comparing the difference between spirit level bubble positions.

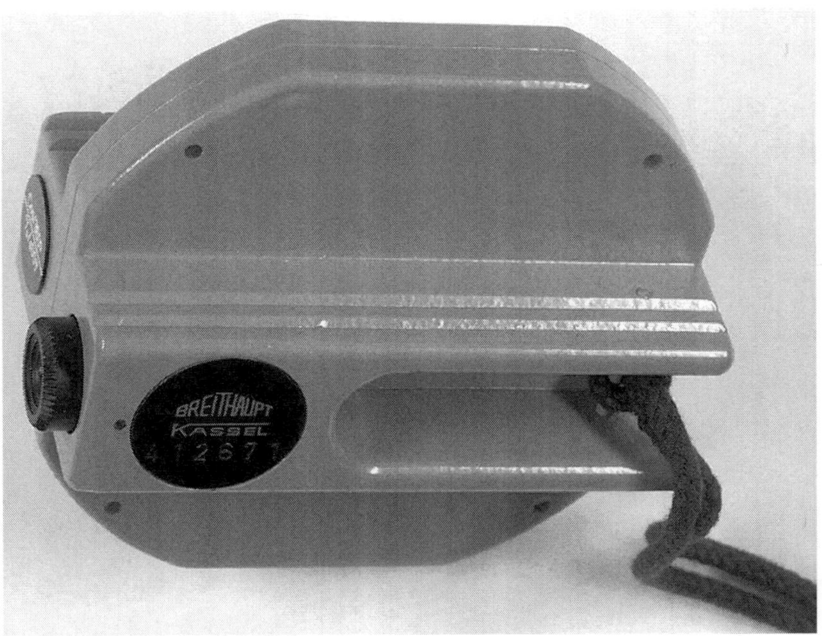

Figure: *Leveler for use by hand*

An automatic level, self-levelling level or builder's auto level, includes an internal compensator mechanism (a swinging prism) that, when set close to level, automatically removes any remaining variation from level.

This reduces the need to set the instrument truly level, as with a dumpy or tilting level. Self-levelling instruments are the preferred instrument on building sites, construction and surveying due to ease of use and rapid setup time.

A digital electronic level is also set level on a tripod and reads a bar-coded staff using electronic laser methods.

The height of the staff where the level beam crosses the staff is shown on a digital display.

This type of level removes interpolation of graduation by a person, thus removing a source of error and increasing accuracy. During night time the dumpy level is used along with auto cross laser for accurate scale readings.

The graphometer or semicircle is a surveying instrument used for angle measurements. It consists of a semicircular limb divided into 180 degrees and sometimes subdivided into minutes.

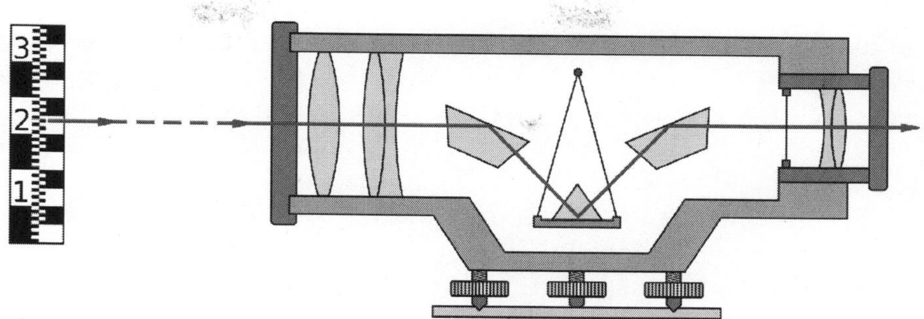

Figure: *An automatic level uses a swinging prism to compensate for small inclination deviations.*

Graphometer

Figure: *Butterfield compass graphometer*

The limb is subtended by the diameter with two sights at its ends. In the middle of the diameter a "box and needle" (a compass) is fixed. On the same middle the alidade with two other sights is fitted. The device is mounted on a staff via a ball and socket joint. In effect the device is a half-circumferentor. For convenience, sometimes another half-circle from 180 to 360 degrees may be graduated in another line on the limb.

The form was introduced in Philippe Danfrie's, *Déclaration de l'usage du graphomètre* (Paris, 1597) and the term "graphometer" was popular with French geodesists. The preferable English-language terms were semicircle or semicircumferentors. Some 19th-century graphometers had telescopic rather than open sights.

Usage

To measure an angle, say, EKG, place the diameter middle C at the angle apex K using the plummet at point C of the instrument. Align the diameter with leg KE of the angle using the sights at the ends of the diameter. Align the alidade with the leg KG using another pair of sights, and read the angle off the limb as marked by the alidade. Further uses of the graphometer are the same as those of the circumferentor.

9

Groma Surveying

The *Groma* or *gruma* was the principal Romansurveying instrument. It comprised a vertical staff with horizontal cross-pieces mounted at right-angles on a bracket. Each cross piece had a plumb line hanging vertically at each end.

Figure: *Groma*

It was used to survey straight lines and right-angles, thence squares or rectangles. The same name was given to:

- the centre of any new military camp, i.e. the point from which was traced the regular grid by using the *groma* instrument
- the centre of a new town, from which the gromatici (surveyors) began to lay out cardo and decumanus grid, with a plough and a pair of oxen

The groma surveying instrument originated from Mesopotamia, from where it was imported by the Greeks in the 4th century BC. Subsequently, it was brought to Rome by the Etruscans.

Gunter's Chain

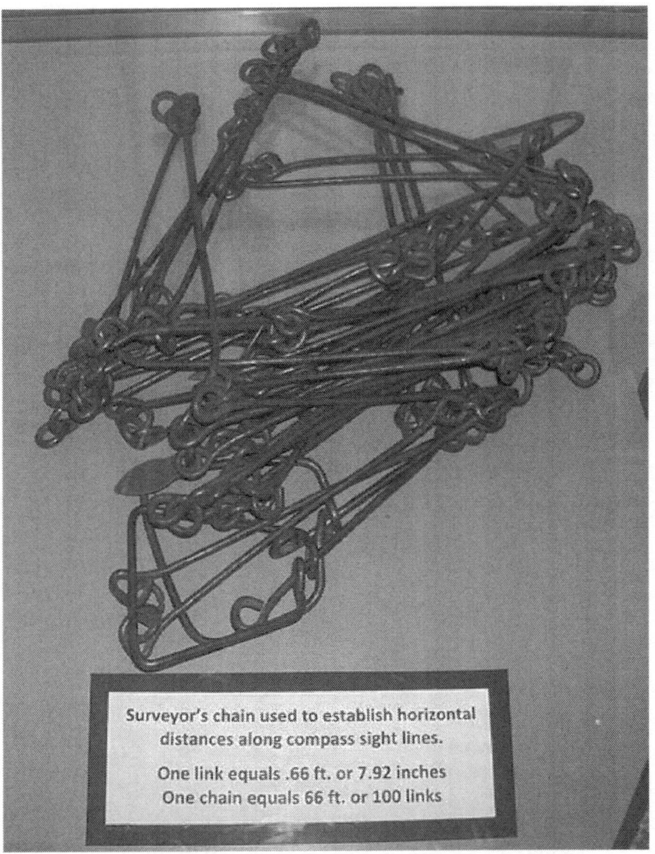

Gunter's chain (also known as Gunter's measurement or Surveyor's measurement) is a geodetic measuring device used for land survey. It was designed and introduced in 1620 by English clergyman and mathematician Edmund Gunter (1581–1626) long before the development of the theodolite and other more sophisticated equipment, enabling plots of land to be accurately surveyed and plotted, for legal and commercial purposes.

Gunter used an actual measuring chain of 100 links. These, the chain and the link, have become units of their own.

Description

The chain is divided into 100 links, marked off into groups of 10 by brass rings which simplify intermediate measurement. Each link is 7.92 inches long, with 10 links making slightly less than 6 feet 8 inches. The full length of the chain is 66 feet. A square link is exactly one hundred-thousandth of an acre and one ten-thousandth of one square chain or 0.0404685642 m^2. It is about 62¾ square inches.

Gunter's chain reconciled two seemingly incompatible systems: the traditional English land measurements, based on the number 4, and the newly introduced system of decimals based on the number 10. Since an acre measured 10 square chains in Gunter's system, the entire process of land measurement could be computed in decimalized chains and links, and then converted to acres by dividing the results by 10.

Method

The method of surveying a field or other parcel of land with Gunter's chain is to first determine corners and other significant locations, and then to measure the distance between them, taking two points at a time. The surveyor is assisted by a chainman. A ranging rod (usually a prominently coloured wooden pole) is placed in the ground at the destination point. Starting at the originating point the chain is laid out towards the ranging rod, and the surveyor then directs the chainman to make the chain perfectly straight and pointing directly at the ranging rod. A pin is put in the ground at the forward end of the chain, and the chain is moved forward so that its hind end is at that point, and the chain is extended again towards the destination point. This process is called ranging, or in the US, chaining; it is repeated until the destination rod is reached, when the surveyor notes how many full lengths (chains) have been laid, and he can then directly read how many links (one-hundredth parts of the chain) are in the distance being measured.

The whole process is repeated for all the other pairs of points required, and it is a simple matter to make a scale diagram of the plot of land. The process is surprisingly accurate and requires only very low technology. Surveying with a chain is simple if the land is level and continuous—it is not physically practicable to range across large depressions or significant waterways, for example. The triangulation method requires that the land is plane (not varying significantly in slope). On sloping land, the chain was to be "leveled" by raising one end as needed, so that undulations did not increase the apparent length of the side or the area of the tract.

Unit of Length

Although Gunter's chain was later superseded by the steel tape, (a form of Tape measure), its legacy was a new unit of length called the chain, which measured 66 feet (or 100 links). This unit still exists as a location identifier on British railways, as well as in some areas of Australia and America. In the United States, for example, Public Lands Survey plats are published in the chain unit to maintain the consistency of a two-hundred-year-old database.

In some places other lengths have been used, for example 8.928 inches (ca. 0.227 m) in Scotland and 10.08 inches (ca. 0.256 m) in Ireland.

The length of a cricket pitch is exactly one chain (22 yards).

Similar Measuring Chains

A similar American system, of lesser popularity, is *Ramsden's* or the *engineer's system*, where the chain consists also of 100 links, each one foot (0.3048 m) long

The even less common *Rathborn system*, also from the 17th century, is based on a 200-link chain of two rods (33 feet, 10.0584 m) length. Each rod (perch) consists of 100 links, (1.98 inches, 50.292 mm each), which are called *seconds* (3), ten of which make a *prime* (2 , 19.8 inches, 0.503 m).

Vincent Wing made chains with 9.90 inch links, most commonly as 33-ft half-chains of 40 links. These chains were sometimes used in the American colonies, particularly Pennsylvania.

Inclinometer

An inclinometer or clinometer is an instrument for measuring angles of slope (or tilt), elevation or depression of an object with respect to gravity. It is also known as a *tilt meter, tilt indicator, slope alert, slope gauge, gradient metre, gradiometre, level gauge, level metre, declinometre,* and *pitch & roll indicator*. Clinometers measure both inclines and declines using three different units of measure: degrees, percent, and topo. Astrolabes are inclinometers that were used for navigation and locating astronomical objects.

In aircraft, the "ball" in turn coordinators or turn and bank indicators is sometimes referred to as an inclinometer.

Figure: *Compass with inclinometer*

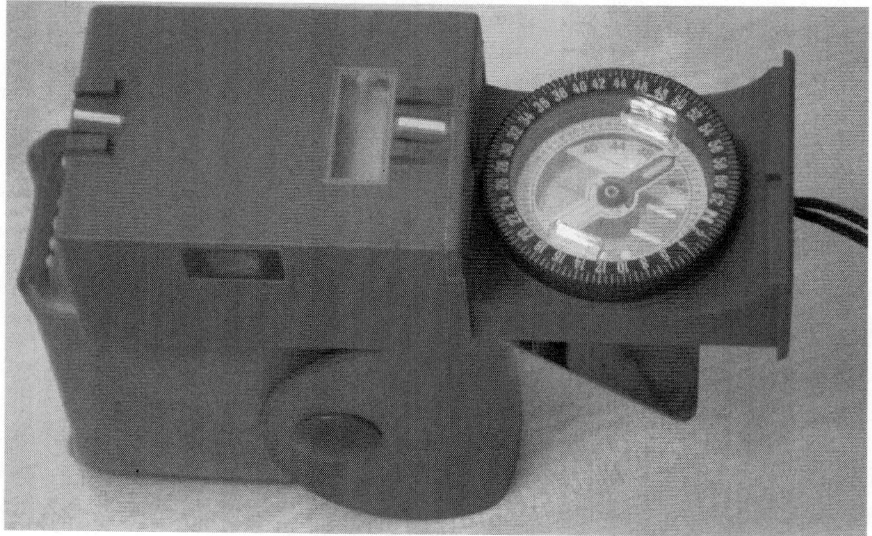

Figure: *Military model*

History

Early inclinometers include examples such as Well's inclinometer, the essential parts of which are a flat side, or base, on which it stands, and a hollow disk just half filled with some heavy liquid. The glass face of the disk is surrounded by a graduated scale that marks the angle at which the surface of the liquid stands, with reference to the flat base.

The zero line is parallel to the base, and when the liquid stands on that line, the flat side is horizontal; the 90 degree is perpendicular to the base, and when the liquid stands on that line, the flat side is perpendicular or plumb. Intervening angles are marked, and, with the aid of simple conversion tables, the instrument indicates the rate of fall per set distance of horizontal measurement, and set distance of the sloping line.

One of the more famous inclinometer installations was on the panel of the Ryan NYP "The Spirit of St. Louis" - in 1927 Charles Lindbergh chose the lightweight Rieker Inc P-1057 Degree Inclinometer to give him climb and descent angle information.

Use

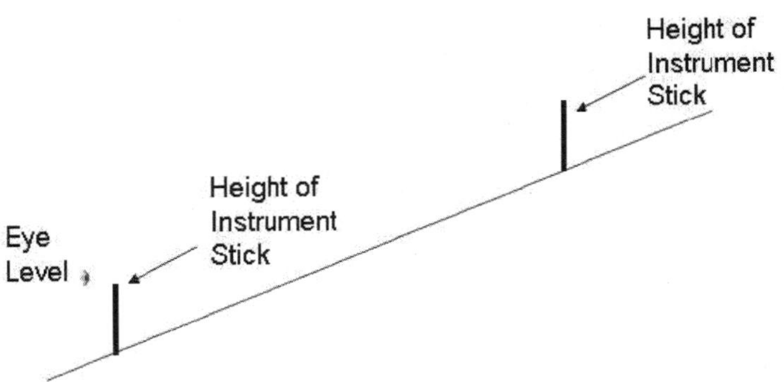

Figure: *Measuring slope with a clinometer*

A forester using a clinometer makes use of basic trigonometry. First the observer measures a straight-line distance D from some observation point O to the object. Then, using the clinometer, the observer measures the angle a between O and the top of the object. Then the observer does the same for the angle b between O and the bottom of the object. Multiplying D by the tangent of a gives the height of the object above the observer, and by the tangent of b the depth of the object below the observer. Adding the two of course gives the total height (H) of the object.

The forester stands at a fixed distance from the base of the tree. The most common distances in the United States are 50 feet (15.24 m), 66 feet (20.12 m), and 100 feet (30.48 m). To obtain accurate readings it is best to use taped measured distance instead of paced distances. For the most accurate readings it is best to use a distance that is not less than the height of the tree being measured., that is, that the clinometer will measure an angle less than 45° (100%).

The observer sights to the top of tree, if total height is the desired measurement. If the desired measurement is merchantable height – that is, the height producing timber that can be sold – the observer sights to a point on the tree above which no more merchantable timber is found. Note that the observer should use slopes expressed as a percent. The observer then measures to the bottom of the tree, again using percents. The observer than takes the slope to the top of the tree, and subtracts the slope of the bottom of the tree from it. Note that when the ground slopes downwards, the observer will record a negative slope to the base of the tree. When this is subtracted from the slope to the top of the tree, it is actually being added. These numbers are then divided by 100, and multiplied by the distance from the tree.

There is still some error in this number, because the distance is measured along the ground, and is not the actual horizontal distance. By standing at the base of the tree, the observer should mark his/ her eye level. They then stand back, and find the slope expressed as a degree to the mark. This is the slope of the ground. By taking the cosine of that number, and multiplying it with the previous number, the observer will calculate the height of the tree. The observer must always measure a leaning tree so that the tree is leaning to the left or right. Measurements should never be taken with the tree leaning toward or away from the observer because this will affect their accuracy because of foreshortening.

The clinometer is also commonly used by foresters to obtain the percent slope of terrain. This measurement is based on the same trigonometric principles described above. Slope measurements, however, require that both observer and target be a constant height above the ground; thus a range pole or height of measurement (HI) stick is often used in slope measurements.

Accuracy

Certain highly sensitive electronic inclinometer sensors can achieve an output resolution to 0.0001 degrees - depending on the technology and angle range, it may be limited to 0.01°. An inclinometer sensor's true or absolute accuracy (which is the combined total error), however, is a combination of initial sets of sensor zero offset and sensitivity, sensor linearity, hysteresis, repeatability, and the temperature drifts of zero and sensitivity - electronic inclinometers accuracy can typically range from .01° to ±2° depending on the sensor and situation. Typically in room ambient conditions the accuracy is limited to the sensor linearity specification.

Figure: *Clinometer designed to enable indirect fire capability with a Vickers machine gun circa 1918*

Sensor Technology

Tilt sensors and inclinometers generate an artificial horizon and measure angular tilt with respect to this horizon. They are used in cameras, aircraft flight controls, automobile security systems, and speciality switches and are also used for platform levelling, boom angle indication, and in other applications requiring measurement of tilt.

Important specifications to consider when searching for tilt sensors and inclinometers are the tilt angle range and number of axes (which are usually, but not always, orthogonal). The tilt angle range is the range of desired linear output.

Common implementations of tilt sensors and inclinometers are accelerometer, Liquid Capacitive, electrolytic, gas bubble in liquid, and pendulum.

Tilt sensor technology has also been implemented in video games. *Yoshi's Universal Gravitation* and *Kirby Tilt 'n' Tumble* are both built around a tilt sensor mechanism, which is built into the cartridge. The PlayStation 3 and Wii game controllers also use tilt as a means to play video games.

Inclinometers are also used in civil engineering, for example to measure the inclination of land to be built upon.

Some inclinometers provide an electronic interface based on CAN (Controller Area Network). In addition, those inclinometers may support the standardized CANopen profile (CiA 410). In this case, these inclinometers are compatible and partly interchangeable.

Specific Functions

Inclinometers are used for:

- Determining latitude using Polaris (in the Northern Hemisphere) or the two stars of the constellation Crux (in the Southern Hemisphere).
- Determining the angle of the Earth's magnetic field with respect to the horizontal plane.
- Showing a deviation from the true vertical or horizontal.
- Surveying, to measure an angle of inclination or elevation.
- Alerting an equipment operator that it may tip over.
- Measuring angles of elevation, slope, or incline, e.g. of an embankment.
- Measuring slight differences in slopes, particularly for geophysics. Such inclinometers are, for instance, used for monitoring volcanoes, or for measuring the depth and rate of landslide movement.
- Measuring movements in walls or the ground in civil engineering projects.
- Determining the dip of beds or strata, or the slope of an embankment or cutting; a kind of plumb level.
- Some automotive safety systems.
- Indicating pitch and roll of vehicles, nautical craft, and aircraft.

- Monitoring the boom angle of cranes and material handlers.
- Measuring the "look angle" of a satellite antenna towards a satellite.
- Adjusting a solar panel to the optimal angle to maximize its output.
- Measuring the slope angle of a tape or chain during distance measurement.
- Measuring the height of a building, tree, or other feature using a vertical angle and a distance (determined by taping or pacing), using trigonometry.
- Measuring the angle of drilling in well logging.
- Measuring the list of a ship in still water and the roll in rough water.
- Measuring steepness of a ski slope.
- Measuring the orientation of planes and lineations in rocks, in combination with a compass, in structural geology.
- Measuring Range of Motion in the joints of the body
- Measuring the inclination angle of the pelvis. Numerous neck and back measurements require the simultaneous use of 2 inclinometers.
- Measuring the angles of elevation to, and ultimately computing the altitudes of, many things otherwise inaccessible for direct measurement.
- Measuring and fine tuning the angle of line array speaker hangs. Confirmation of the angle achieved via use of a laser built into the remote inclinometer.

Laser Line Level

A laser line level is a tool combining a spirit level and/or plumb bob with a laser to display an accurately horizontal or vertical illuminated line on a surface the laser line level is laid against. Laser line levels are used wherever accurate verticals and horizontals are required, typically in the construction and cabinetry industries. Some models are inexpensive enough for do-it-yourself applications.

The laser beam is fanned to produce a thin plane beam accurately horizontal or vertical, rather than a pinpoint beam. The axis of the laser is offset from the wall, so that a pinpoint beam would be parallel to and offset from the wall, and would not illuminate it; the fanned beam will intersect the wall, creating an accurately horizontal (or vertical) illuminated line along it.

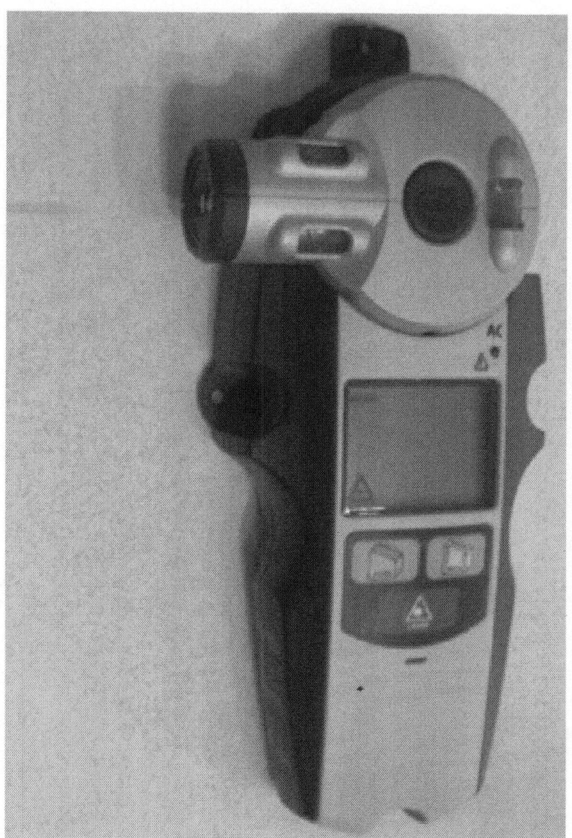

Figure: *Typical consumer laser line level using spirit levels for three planes and including a digital stud sensor display.*

The machine is set up using the built-in spirit level or plumb bob, and the line along the surface is then guaranteed to be accurately horizontal or vertical to within a certain tolerance, specified either in millimetres per metre or fractions of an inch over a specified distance in feet. A more advanced device may be accurate to within 0.3 mm/m; while lower-end models may be closer to 1.5 mm/m. The illuminated line is necessarily absolutely straight, so that the line level can be used as a straightedge; for example, to see if a shelf is warped, even if not horizontal.

Laser Scanning

In modern engineering, the term 'laser scanning' is used with two related, but separate meanings. The first, more general, meaning is the controlled deflection of laser beams, visible or invisible. Scanned laser beams are used in stereolithography machines, in rapid prototyping, in machines for material processing, in laser engraving machines, in ophtalmological laser systems for the treatment of presbyopia, in confocal microscopy, in laser printers, in laser shows, in Laser TV, in LIDAR, and in barcode scanners.

The second, more specific, meaning is the controlled steering of laser beams followed by a distance measurement at every pointing direction.

This method, often called 3D object scanning or 3D laser scanning, is used to rapidly capture shapes of objects, buildings and landscapes.

This article focuses on the general meaning, i.e., on the methods and applications of scanned laser beams.

Technology

Scanning Mirrors

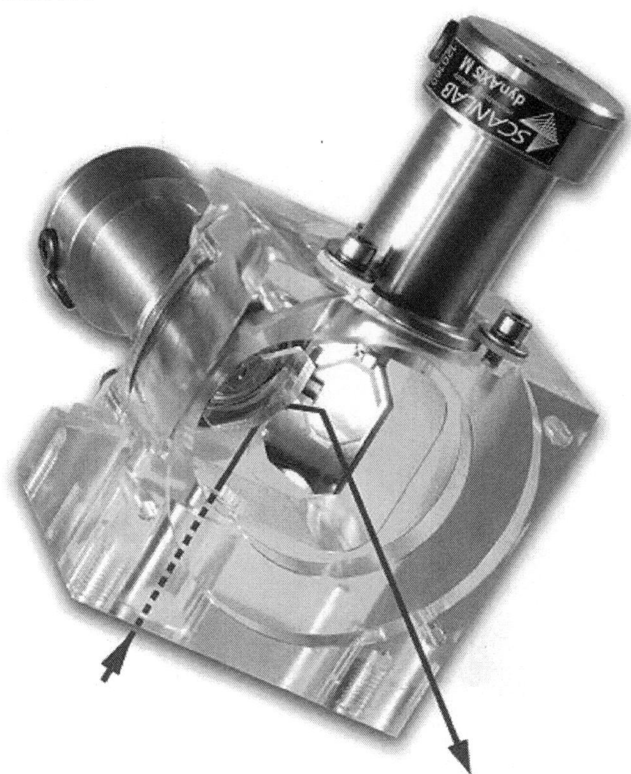

Figure: *Laser scanning module with two galvanometers, from Scanlab AG. The red arrow shows the path of the laser beam.*

Most laser scanners use moveable mirrors to steer the laser beam. The steering of the beam can be one-dimensional, as inside a laser printer, or two-dimensional, as in a laser show system.

Additionally, the mirrors can lead to a periodic motion - like the rotating *mirror polygons* in a barcode scanner or so-called *resonant galvanometer* scanners - or to anfreely addressable motion, as in servo-controlled galvanometer scanners. One also uses the terms raster scanning and vector scanning to distinguish the two situations.

To control the scanning motion, scanners need a rotary encoder and control electronics that provides, for a desired angle or phase, the suitable electrical current to the motor or galvanometer. A software systems usually controls the scanning motion and, if 3D scanning is implemented, also the collection of the measured data.

In order to position a laser beam in two dimensions, it is possible either to rotate one mirror along two axes - used mainly for slow scanning systems - or to reflect the laser beam onto two closely spaced mirrors that are mounted on orthogonal axes. Each of the two flat or polygonal mirrors is then driven by a galvanometer or by an electric motor. Two-dimensional systems are essential for most applications in material processing, confocal microscopy, and medical science.

Some applications require to position the focus of a laser beam in three dimensions. This is achieved by a servo-controlled lens system, usually called a 'focus shifter' or 'z-shifter'.

Many laser scanners further allow changing the laser intensity.

In laser projectors for laser TV or laser displays, the three fundamental colours red blue and green are combined in a single beam and then reflected together over the two mirrors.

The most common way to move mirrors is, as mentioned, the use of an electric motor or of a galvanometer. However piezoelectric actuators or magnetostrictive actuators are alternative options. They offer higher achievable angular speeds, but often at the expense of smaller achievable maximum angles.

Scanning Refractive Optics

When two Risley prisms are rotated against each other, a beam of light can be scanned at will inside a cone. Such scanners are used for tracking missiles.

When two optical lenses are moved or rotated against each other, a laser beam can be scanned in a way similar to mirror scanners.

Material Effects

Some special laser scanners use, instead of moving mirrors, acousto-optic deflectors or electrooptic deflectors. These mechanisms allow the highest scanning frequencies possible so far. They are used, for example, in laser TV systems. On the other hand, these systems are also much more expensive than mirror scanning systems.

Phased Array Scanning

Research is going on to achieve scanning of laser beams through phased arrays. This method is used to scan RADAR beams without moving parts. With the use of Vertical-cavity surface-emitting laser (VCSELs), it might be possible to realize fast laser scanners in the foreseeable future.

Applications

3D Object Scanning

3D object scanning allows enhancing the design process, speeds up and reduces data collection errors, saves time and money, and thus makes it an attractive alternative to traditional data collection techniques. 3D scanning is also used for mobile mapping, surveying, scanning of buildings and building interiors, and in archaeology.

Material Processing

Depending on the power of the laser, its influence on a working piece differs: lower power values are used for laser engraving and laser ablation, where material is partially removed by the laser. With higher powers the material becomes fluid and laser welding can be realized, or if the power is high enough to remove the material completely, then laser cutting can be performed. Modern lasers can cut steel blocks with a thickness of 10 cm and more or ablate a layer of the cornea that is only a few micrometers thick.

The ability of lasers to harden liquid polymers, together with laser scanners, is used in rapid prototyping, the ability to melt polymers and metals is, with laser scanners, to produce parts by laser sintering.

The principle that is used for all these applications is the same: software that runs on a PC or an embedded system and that controls the complete process is connected with a scanner card. That card converts the received vector data to movement information which is sent to the scanhead. This scanhead consists of two mirrors that are able to deflect the laser beam in one level (X- and Y-coordinate). The third dimension is - if necessary - realized by a specific optic that is able to move the laser's focal point in the depth-direction (Z-axis).

Scanning the laser focus in the third spatial dimension is needed for some special applications like the laser scribing of curved surfaces or for in-glass-marking where the laser has to influence the material at specific positions within it. For these cases it is important that the laser has as small a focal point as possible.

For enhanced laser scanning applications and/or high material throughput during production, scanning systems with more than one scanhead are used. Here the software has to control what is done exactly within such a multihead application: it is possible that all available heads have to mark the same to finish processing faster or that the heads mark one single job in parallel where every scanhead performs a part of the job in case of large working areas.

Barcode Readers

Many barcode readers, especially those with the ability to read bar codes at a distance of a few metres, use scanned laser beams. In these devices, a semiconductor laser beam is usually scanned with the help of a resonant mirror scanner. The mirror is driven electromagnetically and is made of a metal-coated polymer.

Space Flight

When a space transporter has to dock to the space station, it must carefully maneuver to the correct position. In order to determine its relative position to the space station, laser scanners built into the front of the space transporter scan the shape of the space station and then determine, through a computer, the maneuvering commands. Resonant galvanometer scanners are used for this application.

Laser Tracker

Laser trackers are instruments that accurately measure large objects by determining the positions of optical targets held against those objects. The accuracy of laser trackers is of the order of 0.025 mm over a distance of several metres. Some examples of laser tracker applications are to align aircraft wings during assembly and to align large machine tools. To take measurements the technician first sets up a laser tracker on a tripod with an unobstructed view of the object to be measured. The technician removes a target from the base of the laser tracker and carries it to the object to be measured, moving smoothly to allow the laser tracker to follow the movement of the target. The technician places the target against the object and triggers measurements to be taken at selected points, sometimes by a remote control device. Measurements can be imported into different types of software to plot the points or to calculate deviation from the correct position. The targets are known as "retroreflective" because they reflect the laser beam back in the same direction it came from (in this case, back to the laser tracker). One type of target in common use is called a spherically mounted retroreflector (SMR), which resembles a ball bearing with mirrored surfaces cut into it.

Macrometer

A macrometer is an instrument for measuring the size and distance of distant objects. Distant in this sense means a length that can not be readily measured by a calibrated length. The optical version of this instrument used two mirrors on a common sextant. By aligning the object on the mirrors using a precise vernier, the position of the mirrors could be used to compute the range to the object. The distance and the angular size of the object would then yield the actual size.

The *Macrometer Interferometric Surveyor* is a commercial GPS-based system for performing geodetic measurements.

Pentaprism

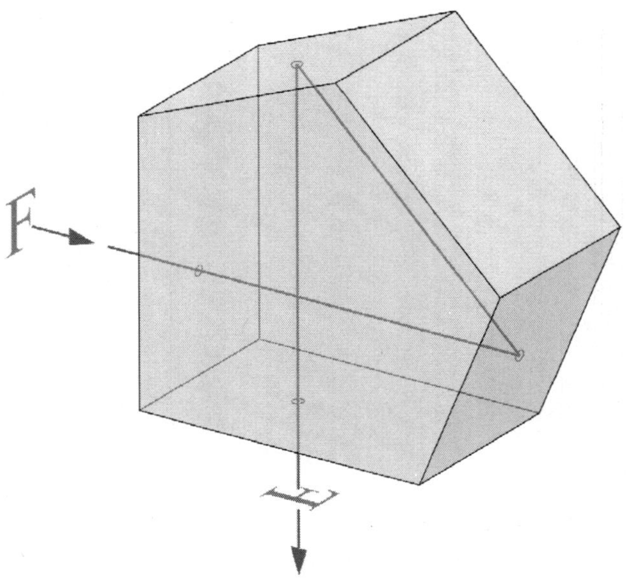

Figure: *A pentaprism.*

A pentaprism is a five-sided reflecting prism used to deviate a beam of light by a constant 90°, even if the entry beam is not at 90° to the prism. The beam reflects inside the prism *twice*, allowing the transmission of an image through a right angle without inverting it (that is, without changing the image's handedness) as an ordinary right-angle prism or mirror would.

The reflections inside the prism are not caused by total internal reflection, since the beams are incident at an angle less than the critical angle (the minimum angle for total internal reflection). Instead, the two faces are coated to provide mirror surfaces. The two opposite transmitting faces are often coated with an antireflection coating to reduce spurious reflections. The fifth face of the prism is not used optically but truncates what would otherwise be an awkward angle joining the two mirrored faces.

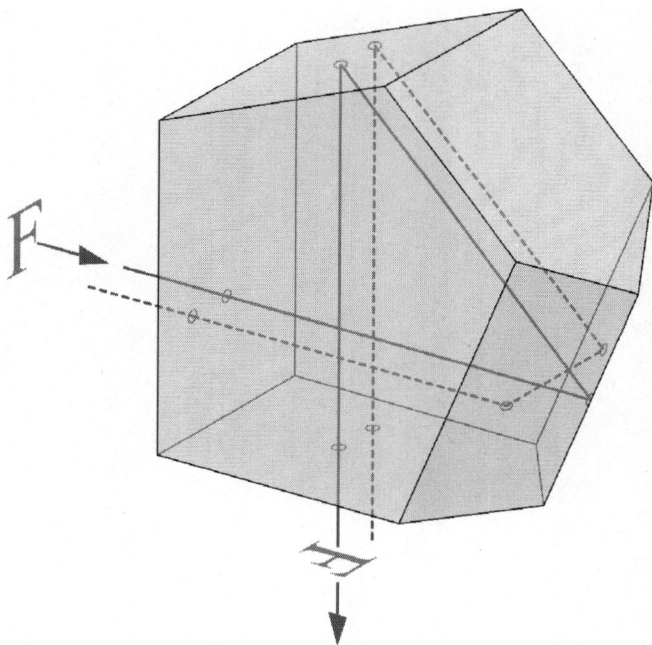

Figure: *A perspective drawing showing a roof pentaprism commonly used in a single lens reflex camera (SLR). The image is flipped laterally by the prism.*

Roof Pentaprism

A variant of this prism is the roof pentaprism which is commonly used in the viewfinder of single-lens reflex cameras. The camera lens renders an image that is both vertically and laterally reversed, and the reflex mirror re-inverts it leaving an image laterally reversed. In this case, the image needs to be reflected left-to-right as the prism transmits the image formed on the camera's focusing screen. This lateral inversion is done by replacing one of the reflective faces of a normal pentaprism with a "roof" section, with two additional surfaces angled towards each other and meeting at 90°, which laterally reverses the image back to normal. Reflex cameras with so-called 'waist level' finders, including many medium format cameras, display a laterally reverse.

Figure: *An image as seen through a roof pentaprism. This is looking in through the eyepiece plane.*

Philadelphia Rod

A Philadelphia rod is a tool used in surveying. The rod is used in levelling procedures to determine elevations. It is read using a level.

A Philadelphia Level Rod consists of two sliding sections graduated in hundredths of a foot. On the front of the rod the graduation increasing from zero at the bottom. On the back of the rod the graduation decrease from 13.09 ft at the bottom to 7 ft. The division of the device in two sliding sections are devised for ease of support.

Readings of 7 ft or less, and up to 13 ft can be measured. It has a rear section that slides on the front section. The rod must be fully extended, when higher measurements are needed to avoid reading errors. Distances of up to 250' may be read.

The rod may be equipped with a target to increase the readable range of the rod. When the target is equipped with a Vernier scale measurements to the thousands of a foot are possible. For readings less than 7 ft the target is attached on the bottom section of the rod and adjust by signals from level operator until the target is inline with the level's horizontal cross hair. For readings greater than 7 ft the target is attached to top section of the rod and the top section is raised/ lowered until it the target intersects with the cross hair of the level. The rod is then locked and the zero of Vernier scale on the back of the rod will be aligned with the target's height.

Plumb Bob

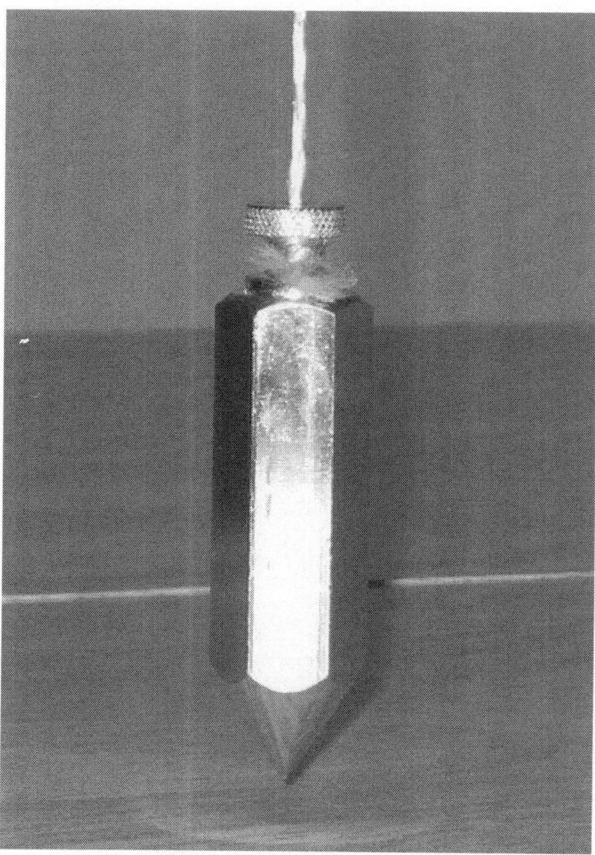

Figure: *A plumb-bob*

A plumb-bob or a plummet is a weight, usually with a pointed tip on the bottom, that is suspended from a string and used as a vertical reference line, or plumb-line. It is essentially the y-axis equivalent of a "water level".

The instrument has been used since at least the time of ancient Egypt to ensure that constructions are "plumb", or vertical. It is also used in surveying to establish the nadir with respect to gravity of a point in space. They are used with a variety of instruments (including levels, theodolites, and steel tapes) to set the instrument exactly over a fixed survey marker, or to transcribe positions onto the ground for placing a marker.

Etymology

The "plumb" in "plumb-bob" comes from the fact that such tools were originally made of lead (Latin*plumbum*, French *plomb*). The adjective "plumb" developed by extension, as did the noun "aplomb", from the notion of "standing upright".

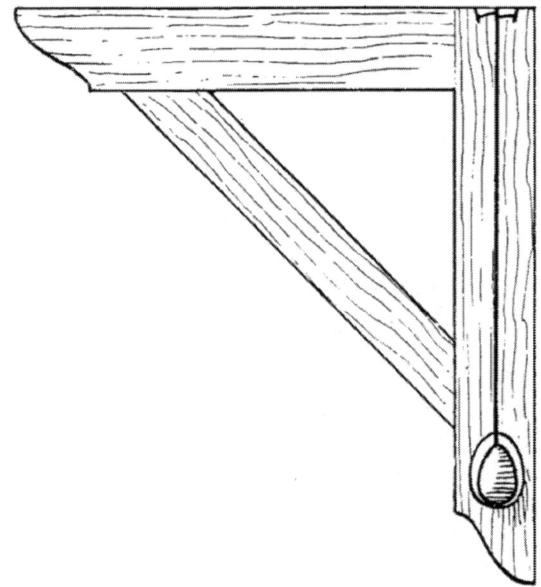

Figure: *A plumb square from the book Cassells' Carpentry and Joinery*

Use

Up until the modern age, on most tall structures, plumb-bobs were used to provide vertical datum lines for the building measurements. A section of the scaffolding would hold a plumb line that was centred over a datum mark on the floor. As the building proceeded upwards the plumb line would also be taken higher, still centred on the datum. Many cathedral spires, domes and towers still have brass datum marks inlaid into their floors, that signify the centre of the structure above.

Although a plumb-bob and line alone can only determine a vertical, if mounted on a suitable scale the instrument may also be used as an inclinometer to measure angles to the vertical.

The early skyscrapers used heavy plumb-bobs hung on wire in their elevator shafts. The weight would hang in a container of oil to dampen any swinging movement, functioning as a shock absorber.

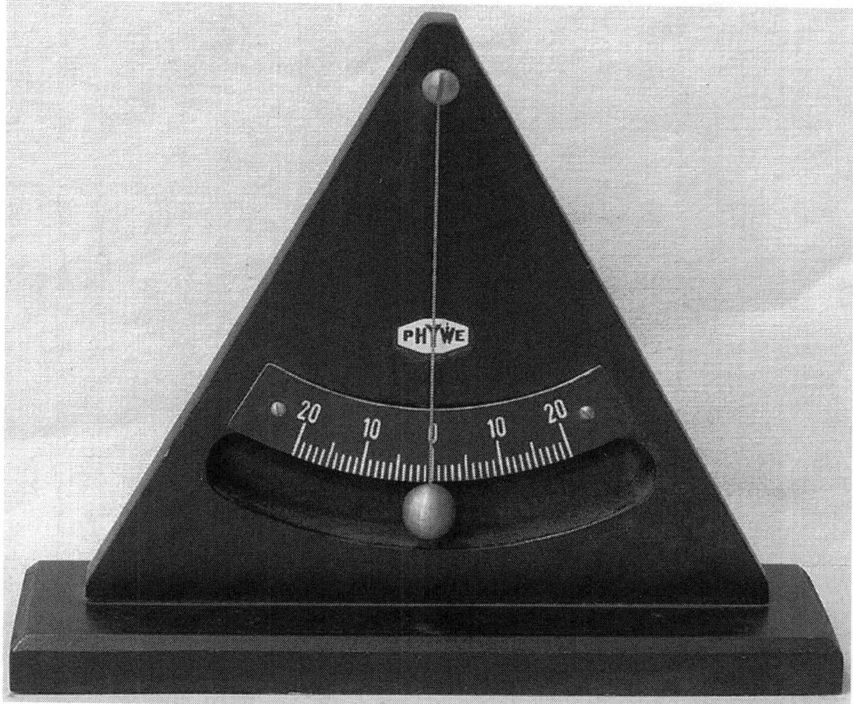

Figure: *Plumb-bob with scale as an inclinometer*

Determining Centre of Gravity of an Irregular Shape

Students of figure drawing will also make use of a plumb line to find the vertical axis through the centre of gravity of their subject and lay it down on paper as a point of reference. The device used may be purpose-made plumb lines, or simply makeshift devices made from a piece of string and a weighted object, such as a metal washer. This plumb line is important for lining up anatomical geometries and visualizing the subject's centre of balance.

Ramsden Theodolite

Figure: *The Great Ramsden Theodolite is now in the Science Museum in London, England.*

The Ramsden theodolite is a large theodolite that was specially constructed for use in the first Ordnance Survey of Southern Britain. It was also known as the Great or 36 inch theodolite.

The theodolite was commissioned from Jesse Ramsden, a leading Yorkshire instrument maker, who had developed an accurate dividing engine for graduatingangular scales. The instrument was accurate to within a second of arc. The theodolite took three years to build and had a base circle of 3 ft (914 mm).

The full survey, sometimes called the Principal Triangulation of Great Britain, was begun in 1791 by a team formed under General William Roy (d 1790). The survey used the new theodolite on a specially surveyed baseline based on Roy's accurate surveys between London and Paris. Traces of the theodolite support structure were still to be found many years afterwards at some remote survey points, such as at Soldiers' Lump, the summit of Black Hill in the Peak District of England. The theodolite is now in the Science Museum in London. A total of 8 such instruments were manufactured to this design and found use as far away as India and Switzerland.

Ramsden, who was elected to the Royal Society in 1786, and was awarded the Copley Medal in 1795 for his instruments, also made important contributions to fields such as optics (the Ramsden eyepiece) and electrostatics (the Ramsden machine).

Repeating Circle

The repeating circle is an instrument for geodetic surveying, invented by Etienne Lenoir in 1784, while an assistant of Jean-Charles de Borda, who later improved the instrument. It was notable as being the equal of the great theodolite created by the renowned instrument maker, Jesse Ramsden. It was used to measure the meridian arc from Dunkirk to Barcelona by Delambre and Méchain.

The repeating circle is made of two telescopes mounted on a shared axis with scales to measure the angle between the two. The instrument combines multiple measurements to increase accuracy with the following procedure:

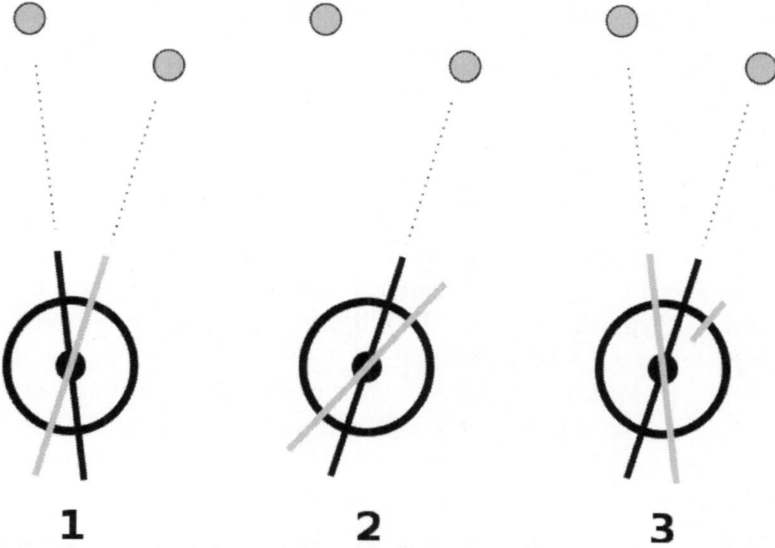

Figure: *Align the instrument so its plane includes the two points to be measured, and aim each telescope at a point. Keeping the angle between the telescopes locked, rotate the left (black) telescope clockwise to aim at the right point. Note the position of the right (gray) telescope, and rotate it back to the left point.*

At this stage, the angle on the instrument is double the angle of interest between the points. Repeating the procedure causes the instrument to show 4x the angle of interest with further iterations increase it to 6x, 8x, and so on. In this way, many measurements can be added together, allowing some of the random measurement errors to cancel out.

Military Sketching Board

The Military Sketching Board was designed to be used on horseback. The board incorporates a compass, an inclinometer, a ruler, a roll of paper and an arm buckle.

Description

The board was designed to be strapped to the arm of a cavalryman on the arm that he held the horses bridle with. The board is attached to the leather buckle with a swivel joint. This means that the user can twist the whole board on their arm to ensure that the compass alignment is clear. The board incorporates a compass for this purpose. The user can then use the supplied ruler. In earlier versions this was attached with rubber bands but later versions had the ruler permanently attached. The paper was about two or three feet long and was rolled around one of the side rollers. As a sketch was completed then the paper roll was advanced to supply more clear paper. On the back of the board is an Inclinometer to allow the severity of ascents to be recorded.

History

The board was originally designed by Colonel W H Richards, who taught military surveying at the Royal Military Academy Sandhurst in about 1880 before he went on to teach in India. The board was improved by Major (later Colonel) William Willoughby Cole Vernerwho had briefly Professor of Topography at the Sandhurst. He patented improvements in 1887 and 1891 and the board became known as *Verner's*. Verner published his own guide to the sketching board in 1889. However the board was not universally loved and some referred to it as *"The damnable cavalry sketching board"*.

In the 1930s very similar devices were used by solo aviators. The device was strapped to the arm or leg and was loaded with maps rather than sketching paper.

One of these devices is in the collection of the National Museums of Scotland.

Surveyor's Wheel

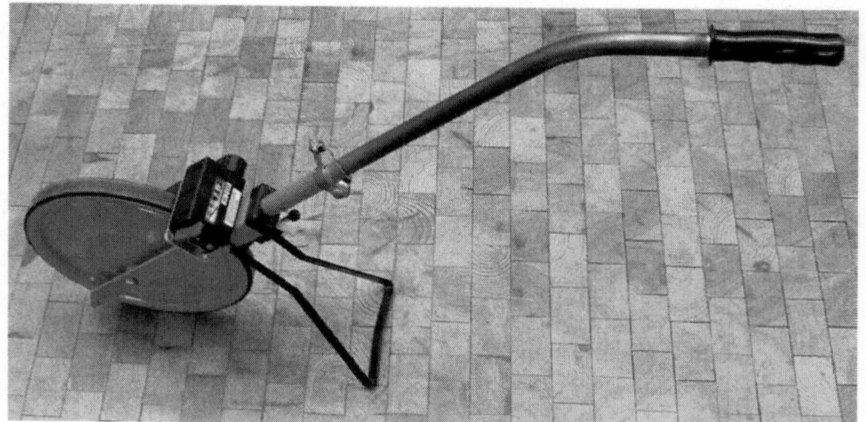

Figure: *Surveyor's wheel*

A surveyor's wheel, also called a clickwheel, hodometer, waywiser, trundle wheel, measuring wheel or perambulator is a device for measuring distance.

Origin

The origins of the surveyor's wheel are connected to the origins of the odometer. While the latter is derived to measure distances travelled by a vehicle, the former is specialized to measure distances. Much of the material on the earliest stages in the development of the hodometer are adequately covered in odometer.

In the 17th century, the surveyor's wheel was re-introduced and used to measure distances. A single wheel is attached to a handle and the device can be pushed or pulled along by a person walking. Early devices were made of wood and may have an iron rim to provide strength.

The wheels themselves would be made in the same manner as wagon wheels and often by the same makers. The measuring devices would be made by makers of scientific instruments and the device and handles would be attached to the wheel by them. The device to read the distance travelled would be mounted either near the hub of the wheel or at the top of the handle.

In some cases, double-wheel hodometers were constructed.

Modern surveyor's wheels are constructed primarily of aluminium, with solid or pneumatictires on the wheel. Some can fold for transport or storage.

Usage of the Surveyor's Wheel

Each revolution of the wheel measures a specific distance, such as a yard, metre or half-rod. Thus counting revolutions with a mechanical device attached to the wheel measures the distance directly.

Surveyor's wheels will provide a measure of good accuracy on a smooth surface, such as pavement. On rough terrain, wheel slippage and bouncing can reduce the accuracy. Soft sandy or muddy soil can also affect the rolling of the wheel. As well, obstacles in the way of the path may have to be accounted for separately. Good surveyors will keep track of any circumstance on the path that can influence the accuracy of the distance measured and either measure that portion with an alternative, such as a surveyor's tape or measuring tape, or make a reasonable estimate of the correction to apply.

Surveyor's wheels are used primarily for lower accuracy surveys. They are often used by road maintenance or underground utility workers and by farmers for fast measures over distances too inconvenient to measure with a surveyor's tape.

The surveyor's wheel measures the distance along a surface, whereas in normal land surveying, distances between points are usually measured horizontally with vertical measurements indicated in differences in elevation. Thus conventionally surveyed distances will be less than those measured by a surveyor's wheel.

Topographic Abney Level

A Topographic Abney Level is an instrument used in surveying which consists of a fixed sighting tube, a movable spirit level that is connected to a pointing arm, and a protractor scale. The Topographic Abney Level is an easy to use, relatively inexpensive, and, when used correctly, an accurate surveying tool. The Topographic Abney Level is used to measure degrees, percent of grade, topographic elevation, and chainage correction.

By using trigonometry the user of a Topographic Abney Level can determine height, volume, and grade. The Topographic Abney Level is used at the eye height of the surveyor and is best employed when teamed with a second surveyor of the same eye height. This allows for easy sighting of the level and greater accuracy. A ranging pole can be marked at the eye height of the level user or the approximate location of the eye height (i.e. chin, nose, top of head) of the level surveyor must be known of the ranging surveyor.

Origins

The Abney level was invented by Sir William de Wiveleslie Abney (Born 24 Jul 1843 Died 3 Dec 1920) who was an English astronomer and chemist best known for his pioneering of colour photography and colour vision. Abney invented this instrument under the employment of the School of Military Engineering in Chatham, England in the 1870s. It is described by W. & L. E. Gurley as an English modification of the Locke hand level, noting that it gives angles of elevation and is also divided for slopes, as 1 to 2, 2 to 1, etc. Since the main tube of this instrument is square, it can be applied to any plane surface. The clinometer scale is graduated to degrees, and read by vernier to 5 minutes.

Usage

Use of an Abney level is discussed in volume 12 of the *Forest Quarterly*[1] published by the New York State College of Forestry in 1914. Discussion on the use of the Abney level starts on page 370.

Total Station

A total station is an electronic/optical instrument used in modern surveying and building construction. The total station is an electronic theodolite (transit) integrated with an electronic distance metre (EDM) to read slope distances from the instrument to a particular point.

Robotic total stations allow the operator to control the instrument from a distance via remote control. This eliminates the need for an assistant staff member as the operator holds the reflector and controls the total station from the observed point.

Technology

Coordinate Measurement

Coordinates of an unknown point relative to a known coordinate can be determined using the total station as long as a direct line of sight can be established between the two points. Angles and distances are measured from the total station to points under survey, and the coordinates (X, Y, and Z or easting, northing and elevation) of surveyed points relative to the total station position are calculated using trigonometry and triangulation. To determine an absolute location a Total Station requires line of sight observations and must be set up over a known point or with line of sight to 2 or more points with known location.

For this reason, some total stations also have a Global Navigation Satellite System receiver and do not require a direct line of sight to determine coordinates. However, GNSS measurements may require longer occupation periods and offer relatively poor accuracy in the vertical axis.

Angle Measurement

Most modern total station instruments measure angles by means of electro-optical scanning of extremely precise digital bar-codes etched on rotating glass cylinders or discs within the instrument. The best quality total stations are capable of measuring angles to 0.5 arc-second. Inexpensive "construction grade" total stations can generally measure angles to 5 or 10 arc-seconds.

Distance Measurement

Measurement of distance is accomplished with a modulated microwave or infrared carrier signal, generated by a small solid-state emitter within the instrument's optical path, and reflected by a prism reflector or the object under survey. The modulation pattern in the returning signal is read and interpreted by the computer in the total station. The distance is determined by emitting and receiving multiple frequencies, and determining the integer number of wavelengths to the target for each frequency. Most total stations use purpose-built glass corner cube prism reflectors for the EDM signal. A typical total station can measure distances with an accuracy of about 1.5 millimetres (0.0049 ft) + 2 parts per million over a distance of up to 1,500 metres (4,900 ft).

Reflectorless total stations can measure distances to any object that is reasonably light in colour, up to a few hundred metres.

Data Processing

Some models include internal electronic data storage to record distance, horizontal angle, and vertical angle measured, while other models are equipped to write these measurements to an external data collector, such as a hand-held computer.

When data is downloaded from a total station onto a computer, application software can be used to compute results and generate a map of the surveyed area. The new generation of total stations can also show the map on the touch-screen of the instrument right after measuring the points.

Applications

Total stations are mainly used by land surveyors and civil engineers, either to record features as in topographic surveying or to set out features (such as roads, houses or boundaries). They are also used by archaeologists to record excavations and by police, crime scene investigators, private accident reconstructionists and insurance companies to take measurements of scenes.

Mining

Total stations are the primary survey instrument used in mining surveying. A total station is used to record the absolute location of the tunnel walls (stopes), ceilings (backs), and floors as the drifts of an underground mine are driven. The recorded data are then downloaded into a CAD programme, and compared to the designed layout of the tunnel.

The survey party installs control stations at regular intervals. These are small steel plugs installed in pairs in holes drilled into walls or the back. For wall stations, two plugs are installed in opposite walls, forming a line perpendicular to the drift. For back stations, two plugs are installed in the back, forming a line parallel to the drift.

A set of plugs can be used to locate the total station set up in a drift or tunnel by processing measurements to the plugs by intersection and resection.

Mechanical and Electrical Construction

Total stations have become the highest standard for most forms of construction layout. It is most often used in the X and Y axis to layout the locations of penetrations out of the underground utilities into the foundation, between floors of a structure, as well as roofing penetrations. Because more commercial and industrial construction jobs have become centred around Building Information Modelling (BIM) the coordinates for virtually every pipe, conduit, duct and hanger support are available with digital precision.

The application of communicating a virtual model to a tangible construction potentially eliminates labour costs related to moving poorly measured systems, as well as time spent laying out these systems in the midst of a full blown construction job in progress.

10

Cave Survey

A cave survey is a map of all or part of a cave system, which may be produced to meet differing standards of accuracy depending on the cave conditions and equipment available underground. Cave surveying and cartography, i.e. the creation of an accurate, detailed map, is one of the most common technical activities undertaken within a cave and is a fundamental part of speleology. Surveys can be used to compare caves to each other by length, depth and volume, may reveal clues on speleogenesis, provide a spatial reference for other areas of scientific study and assist visitors with route-finding.

Traditionally, cave surveys are produced in two-dimensional form due to the confines of print, but given the three-dimensional environment inside a cave, modern techniques using computer aided design are increasingly used to allow a more realistic representation of a cave system.

History

The first known plan of a cave dates from 1546, and was of a man-made cavern in tufa called the Stufe di Nerone (Nero's Oven) in Pozzuoli near Naples in Italy. The first natural cave to be mapped was the Baumannshöhle in Germany, of which a sketch from 1656 survives. Another early survey dates from before 1680, and was made by John Aubrey of Long Hole in the Cheddar Gorge. it consists of an elevational section of the cave. Numerous other surveys of caves were made in the following years, though most are sketches and are limited in accuracy. The first cave that is likely to have been accurately surveyed with instruments is the Grotte de Miremont in France. This was surveyed by a civil engineer in 1765 and includes numerous cross-sections. Édouard-Alfred Martel was the first person to describe surveying technique. His surveys were made by having an assistant walk down the passage until they were almost out of sight. Martel would then take a compass bearing to the assistant's light, and measure the distance by pacing up to the assistant. This would equate to a modern day BCRA Grade 2 survey.

The first cave to have its centreline calculated by a computer is the Fergus River Cave in Ireland, which was plotted by members of the UBSS in 1964. The software was programmeed onto a large university mainframe computer and a paper plot was produced.

Methodology

There are many variations to surveying methodology, but most are based on a similar set of steps which haven't changed fundamentally in 250 years, although the instruments (compass and tape) have got smaller and more accurate. Since the late 1990s digital instruments such as distometres have started to change the process, leading to the advent of fully paperless surveying around 2007. The main variation on the normal methodology detailed below have been devices such as LIDAR and SONAR surveyors that produce a point cloud rather than a series of linked stations. Video-based surveying also exists in prototype form.

Surveying

A survey team begins at a fixed point (such as the cave entrance) and measures a series of consecutive line-of-sight measurements between stations. The stations are temporary fixed locations chosen chiefly for their ease of access and clear sight along the cave passage. In some cases, survey stations may be permanently marked to create a fixed reference point to which to return at a later date.

The measurements taken between the stations include:

- direction (azimuth or bearing) taken with a compass
- inclination from horizontal (dip) taken with a clinometer
- distance measured with a low-stretch tape or laser rangefinder
- optionally, distance to surrounding walls — left, right, up, down (*LRUD*)

Coincident with recording straight-line data, details of passage dimensions, shape, gradual or sudden changes in elevation, the presence or absence of still or flowing water, the location of notable features and the material on the floor are recorded, often by means of a sketch map.

Drawing a Line-plot

Later, the cartographer analyses the recorded data, converting them into two-dimensional measurements by way of geometrical calculations. From them he/she creates a *line-plot*; a scaled geometrical representation of the path through the cave.

Finalising

The cartographer then draws details around the line-plot, using the additional data of passage dimensions, water flow and floor/wall topography recorded at the time, to produce a completed cave survey. Cave surveys drawn on paper are often presented in two-dimensional *plan* and/or *profile* views, while computer surveys may simulate three dimensions. Although primarily designed to be functional, some cavers consider cave surveys as an art form.

Hydrolevelling

Hydrolevelling is an alternative to measuring depth with clinometer and tape that has a long history of use in Russia. The technique is regularly used in building construction for finding two points with the same height, as in levelling a floor. In the simplest case, a tube with both ends open is used, attached to a strip of wood, and the tube is filled with water and the depth at each end marked. In Russia, measuring the depth of caves by hydrolevelling began in the 1970s, and was considered to be the most accurate means of measuring depth despite the difficulties in using the cumbersome equipment of the time. Interest in the method has been revived following the discovery of Voronja on the Arabica Massif in the Caucasus—currently the world's deepest cave.

The hydrolevel device used in recent Voronja expeditions comprises a 50-metre (160 ft) transparent tube filled with water, which is coiled or placed on a reel. A rubber glove which acts as a reservoir is placed on one end of the tube, and a metal box with a transparent window is placed on the other. A diver's digital wristwatch with a depth gauge function is submerged in the box. If the rubber glove is placed on one station and the box with the depth gauge is placed on a lower one, then the hydrostatic pressure between the two points depends only on the difference in heights and the density of the water, i.e. the route of the tube does not affect the pressure in the box. Reading the depth gauge gives the apparent depth change between the higher and lower station. Depth changes are 'apparent' because depth gauges are calibrated for sea water, and the hydrolevel is filled with fresh water. Therefore a coefficient must be determined to convert apparent depth changes to true depth changes. Adding the readings for consecutive pairs of stations gives the total depth of the cave.

Accuracy

The accuracy, or *grade*, of a cave survey is dependent on the methodology of measurement. A common survey grading system is that created by the British Cave Research Association in the 1960s, which uses a scale of six grades.

BCRA Grading System

BCRA Gradings for a Cave Line Survey

Grade: Sketch of low accuracy where no measurements have been made

Grade: May be used, if necessary, to describe a sketch that is intermediate in accuracy between Grade 1 & 3

Grade: A rough magnetic survey. Horizontal & vertical angles measured to ±2.5 °; distances measured to ±50 cm; station position error less than 50 cm.

Grade: May be used, if necessary, to describe a survey that fails to attain all the requirements of Grade 5 but is more accurate than a Grade 3 survey.

Grade: A Magnetic survey. Horizontal and vertical angles measured to ±1 °; distances should be observed and recorded to the nearest centimetre and station positions identified to less than 10 cm.

Grade: A magnetic survey that is more accurate than grade 5.

Grade: A survey that is based primarily on the use of a theodolite or total station instead of a compass.

Notes

1. The above table is a summary and is intended only as an aide memoire; the definitions of the survey grades given above must be read in conjunction with these notes.
2. In all cases it is necessary to follow the spirit of the definition and not just the letter.
3. To attain Grade 3 it is necessary to use a clinometer in passages having appreciable slope.
4. To attain Grade 5 it is essential for instruments to be properly calibrated, and all measurements must be taken from a point within a 10 cm diameter sphere centred on the survey station.

5. A Grade 6 survey requires the compass to be used at the limit of possible accuracy, i.e. accurate to ±0.5 °; clinometer readings must be to the same accuracy. Station position error must be less than ±2.5 cm, which will require the use of tripods at all stations or other fixed station markers ('roofhooks').
6. A Grade X survey must include on the drawing notes descriptions of the instruments and techniques used, together with an estimate of the probable accuracy of the survey compared with Grade 3, 5 or 6 surveys.
7. Grades 2 and 4 are for use only when, at some stage of the survey, physical conditions have prevented the survey from attaining all the requirements for the next higher grade and it is not practical to re-survey.
8. Caving organisations, etc., are encouraged to reproduce Table in their own publications; permission is not required from BCRA to do so, but the tables must not be reprinted without these notes.
9. Grade X is only potentially more accurate than Grade 6. It should never be forgotten that the theodolite/Total Station is a complex precision instrument that requires considerable training and regular practice if serious errors are not to be made through its use!
10. In drawing up, the survey co-ordinates must be calculated and not hand-drawn with scale rule and protractor to obtain Grade 5.

BCRA Gradings for Recording Cave Passage Detail

Class A: All passage details based on memory.

Class B: Passage details estimated and recorded in the cave.

Class C: Measurements of detail made at survey stations only.

Class D: Measurements of detail made at survey stations and wherever else needed to show significant changes in passage dimensions.

Notes

1. The accuracy of the detail should be similar to the accuracy of the line.
2. Normally only one of the following combinations of survey grades should be used:
   - 1A
   - 3B or 3C
   - 5C or 5D

- o 6D
- o XA, XB, XC or XD

Survey Error Detection

The equipment used to undertake a cave survey continues to improve. The use of computers, inertia systems, and electronic distance finders has been proposed, but few practical underground applications have evolved at present. Despite these advances, faulty instruments, imprecise measurements, recording errors or other factors may still result in an inaccurate survey, and these errors are often difficult to detect. Some cave surveyors measure each station twice, recording a *back-sight* to the previous station in the opposite direction. A back-sight compass reading that is different by 180 degrees and a clinometer reading that is the same value but with the reverse direction (positive rather than negative, for example) indicates that the original measurement was accurate. When a loop within a cave is surveyed back to its starting point, the resulting line-plot should also form a closed loop. Any gap between the first and last stations is called a *loop-closure error*. If no single error is apparent, one may assume the loop-closure error is due to cumulative inaccuracies, and cave survey software can 'close the loop' by averaging possible errors throughout the loop stations. Loops to test survey accuracy may also be made by surveying across the surface between multiple entrances to the same cave.

The use of a low-frequency cave radio can also verify survey accuracy. A receiving unit on the surface can pinpoint the depth and location of a transmitter in a cave passage by measurement of the geometry of its radio waves. A survey over the surface from the receiver back to the cave entrance forms an artificial loop with the underground survey, whose loop-closure error can then be determined. In the past, cavers were reluctant to redraw complex cave maps after detecting survey errors. Today, computer cartography can automatically redraw cave maps after data has been corrected.

Surveying Software

There is a large number of surveying packages available on various computer platforms, most of which have been developed by cavers with a basis in computer programming. Many of the packages perform particularly well for specific tasks, and as such many cave surveyors will not solely choose one product over another for all cartographic tasks. A popular programme for producing a centerline survey is Survex, which was originally developed by members of the Cambridge University Caving Club for processing survey data from club expeditions to Austria. It was released to the public in 1992. The centerline data can then be exported in various formats and the cave detail drawn in with various other programmes such as AutoCAD, Adobe Illustrator and Inkscape. Other programmes such as Tunnel and Therion have full centerline and map editing capabilities. Therion notably, when it closes survey loops, warps the passages to fit over their length, meaning that entire passages do not have to be redrawn.

Terrestrial LiDAR units are increasing significantly in accuracy and decreasing in price. Several Caves have been "scanned" using both "time of flight" and "phase shift" LiDAR units. The differences are in the relative accuracies available to each. The Oregon Caves National Park, was LiDAR scanned in August of 2011, as were the Paisley Caves Archaeological dig sigte in SE Oregon. Both were scanned with a FARO Focus Phase shift scanner with +/-2mm accuracy. The Oregon Caves were scanned from the main public entrance to the 110 exit and were loop surveyed to the point of beginning. The data is not yet available for public use, but copies are retained by both the US Park Service and i-TEN Associates in Portland, Oregon.

Automated Methods

In recent years an underground geographic positioning technology called HORTA has been utilized in the mining industry. The technology utilizes a gyroscope and an accelerometer to aid in 3D-position determination. Such automated methods have provided a more than fifty-fold increase in underground surveying productivity with more accurate and finer detail maps as well.

Soil Survey

Soil survey, or soil mapping, is the process of classifying soil types and other soil properties in a given area and geo-encoding such information. It applies the principles of soil science, and draws heavily from geomorphology, theories of soil formation, physical geography, and analysis of vegetation and land use patterns. Primary data for the soil survey are acquired by field sampling and by remote sensing. Remote sensing principally uses aerial photography, but LiDAR and other digital techniques are steadily gaining in popularity. In the past, a soil scientist would take hard-copies of aerial photography, topo-sheets, and mapping keys into the field with them. Today, a growing number of soil scientists bring a ruggedized tablet computer and GPS into the field with them. The tablet may be loaded with digital aerial photos, LiDAR, topography, soil geodatabases, mapping keys, and more. The term *soil survey* may also be used as a noun to describe the published results. In the United States, these surveys were once published in book form for individual counties by the National Cooperative Soil Survey. Today, soil surveys are no longer published in book form; they are published to the web and accessed on NRCS Web Soil Survey where a person can create a custom soil survey. This allows for rapid flow of the latest soil information to the user. In the past it could take years to publish a paper soil survey. Today it takes only moments for changes to go live to the public. Also, the most current soil survey data is made available at NRCS Soil Data Mart for high end GIS users such as professional consulting companies and universities.

The information in a soil survey can be used by farmers and ranchers to help determine whether a particular soil type is suited for crops or livestock and what type of soil management might be required. An architect or engineer might use the engineering properties of a soil to determine whether it is suitable for a certain type of construction. A homeowner may even use the information for maintaining or constructing their garden, yard, or home.

Soil Survey Components

Typical information in a published county soil survey includes the following:

- a brief overview of the county's geography
- a general soil map with a brief description of each of the major soil types found in the county along with their characteristics
- detailed aerial photographs with specific soil types outlined and indexed
- photographs of some of the typical soils found in the area
- tables containing general information about the various soils such as total area, comparisons of production of typical crops and common range plants. They also include extensive interpretations for Land use planning such as limitations for dwellings with and without basements, shallow excavations, small commercial buildings, septic tank adsorptions, suitability for development, construction, and water management.
- tables containing specific physical, chemical, and engineering properties such as soil depth, soil texture, particle size and distribution, plasticity, permeability, available water capacity, shrink-swell potential, corrosion properties, and erodibility.

11

Archaeological Field Survey

Archaeological field survey is a type of field research by which archaeologists (often landscape archaeologists) search for archaeological sites and collect information about the location, distribution and organization of past human cultures across a large area (e.g. typically in excess of one hectare, and often in excess of many km^2). Archaeologists conduct surveys to search for particular archaeological sites or kinds of sites, to detect patterns in the distribution of material culture over regions, to make generalizations or test hypotheses about past cultures, and to assess the risks that development projects will have adverse impacts on archaeological heritage. The surveys may be: (a) *intrusive* or *non-intrusive*, depending on the needs of the survey team (and the risk of destroying archaeological evidence if intrusive methods are used) and; (b) *extensive* or *intensive*, depending on the types of research questions being asked of the landscape in question. Surveys can be a practical way to decide whether or not to carry out an excavation (as a way of recording the basic details of a possible site), but may also be ends in themselves, as they produce important information about past human activities in a regional context.

A common role of field survey is in assessment of potential archaeological significance of places where development is proposed. This is usually connected to construction work and road building. The assessment determines whether the area of development impact is likely to contain significant archaeological resources and makes recommendations as to whether the archaeological remains can be avoided or an excavation is necessary before development work can commence. Archaeologists use a variety of tools in survey, including GIS, GPS, remote sensing, geophysical survey and aerial photography.

Research and Planning

Survey work may be undertaken in response to a specific threat (such as proposed or pending development project) to an area of known or unknown archaeological interest or as part of a programme addressing specific research topics. In either case actual fieldwork is most likely to be preceded by a phase of desktop research (reviewing existing data in the form of maps, formal and informal written records, photographs and drawings) or in the modern age internet research using search engines, ancestry and birth or property records online. Consideration should be given to the nature of the landscape (vegetation coverage, existing settlement or industry, soil depth, climate) before a range of techniques is selected to be applied within an appropriate overarching method.

Rationale

An area may be considered worthy of surveying based on the following:

- Artifacts found: Locals have picked up physical artifacts, sometimes held by the local museum but more often collected in private homes or old buildings such as churches and synagogues, and it is unclear where they are coming from.
- Literary sources: Old literary sources have provided archaeologists with clues about settlement locations that have not been archaeologically documented. Sometimes the texts may be quite recent; for instance, a book on local history may mention an interesting area.
- Oral sources: In many locations, local stories contain some hint of a greater past, and often they have a basis in history. For instance, someone may remember that a grandfather who used to walk the hills as a shepherd used to talk about columns from an old temple, although the descendant never saw the ruins.
- Local knowledge: In many cases, locals know where to find something of interest to archaeologists. They may not have reported it because of taking it as part of their world, or because of fearing intrusions on their land or community.
- Previous surveys: In some places, a past survey may have been recorded in an academic journal. The use of more recent technologies and finds from other sites may provide reason to re-examine the site.

- Previous excavations: Excavations carried out before the middle of the 20th century are notoriously poorly documented. They were often carried out by methods that left behind much of the evidence the modern-day archaeologist is looking for. Early excavators were often interested only in fine pottery, jewellery and statues and referred to as rescue archaeologists.
- Lack of knowledge: Many areas of the world have developed limited knowledge about the nature and organization of past human activity at the regional level. (Although one or more sites may be known from an area, often little is known about the wider distribution of contemporary settlements, and how settlement patterns may change over time.) An archaeological field survey is the primary tool for discovering information about previously uninvestigated areas.
- Archaeological hypotheses: Some kinds of archaeological theories — about changes in agricultural strategies or population density for example — are investigated or tested through the use of archaeological surveys of areas that should or should not contain particular kinds of archaeological materials if the theory is true.

Aerial Photography

Aerial photography is a good tool for planning a survey. Remains of older buildings often show in fields as cropmarks; just below the topsoil, the remains may affect the growth of crops or grass. There should preferably be photographs of the same area at different times of the year, allowing the analyst to find the best time to see cropmarks.

Previous Work in the Region

If the indicator that started the process was not a record of previous work, the archaeologists will need to check if any work has been done prior to commencement of the pending project. As many older surveys and excavations were published in papers that are not widely available, this may be a difficult task. A common way to handle this is through a visit to the area, to check with local museums, historians and older people who might remember something about the former activities in a particular locale.

Permissions

It is usually a simple matter to gain permission to perform a cultural field survey, especially a non-intrusive one. If the area is privately owned, the local laws may or may not require the landowners' co-operation.

Permission for an intrusive form of survey may be more difficult to acquire, due to the fear of destroying evidence or property values and the threat of lawsuit for said damages from the property owner.

Intrusive vs. Non-intrusive Surveys

In a non-intrusive survey, nothing is touched, just recorded. An accurate survey of the earthworks and other features can enable them to be interpreted without the need for excavation. An intrusive survey can mean different things. In some cases, all artifacts of archaeological value are collected. This is often the case if it is a rescue survey, but less common in a regular survey.

Another form of intrusive research is bore holes. Small holes are drilled into the ground, most often with hand-powered bores. The contents are examined to determine the depths at which one might find cultural layers, and where one might expect to strike virgin soil. This can be valuable in determining the cost of an excavation - if there is a build-up of several metres of soil above the layers the archaeologist is interested in, the price will obviously be much higher than if artifacts are found only centimeters below ground.

Extensive vs. Intensive Survey

One way to classify archaeological field surveys is to divide them into two types: *intensive* survey and *extensive* survey. The former is characterised by the complete or near-complete coverage of the survey area at a high-resolution, most often by having teams of survey archaeologists walk in a systematic way (e.g. in parallel transects) over parcels of the landscape in question, documenting archaeological data such as lithics, ceramics and/or building remains. However, variations in artifact visibility related to topography, vegetation, and soil character, not to mention the imperfect detection abilities of human observers, bring into question the very concept of complete coverage.

The Extensive survey, on the other hand, is characterised by a low-resolution approach over targets within a study area (sometimes including hundreds of km^2). Sometimes this involves a random sampling or some other kind of probability sample to gain a repsentative sample of the study area. Extensive surveys may be designed to target the identification of archaeological sites across a large area, whereas intensive surveys are designed to provide a more comprehensive picture of the location of sites and the nature of off-site data (e.g. field systems, isolated finds, etc.). Intensive survey is the more costly, timely, and ultimately informative of the two approaches, although extensive survey can provide important information about previously unknown areas.

Purposive vs. Sampling Survey

Archaeological field surveys can also be characterized as either *purposive* or *sampling* surveys. The former, sometimes also called "archaeological prospection," involves cases where archaeologists are searching for a particular site or a particular kind of archaeological material. For example, they might be searching for a particular shipwreck or an historic fort whose exact location is no longer certain. However, they may also be searching for archaeological materials in particular locations to test hypotheses about past use of those spaces. Sampling surveys, on the other hand, have the goal of obtaining a representative sample of some population of sites or artifacts in order to make generalizations about that population. This involves some probability sampling of spatial units, such as random or stratified random sampling of geometrical (often square) or irregular spatial units.

Fieldwalking (Transects)

Conventionally, fieldwalking in grids or along lines called transects has formed the backbone of archaeological survey fieldwork, at least where visibility is fairly good. A single researcher or team will walk slowly through the target area looking for artifacts or other archaeological indicators on the surface, often recording aspects of the environment at the time. The method works best on either ploughed ground or surfaces with little vegetation. On ploughed surfaces, as the soil is turned regularly artifacts will move to the top. Erosion and soil loss on uncultivated and lightly vegetated soil (e.g., in semi-arid environments) may cause artifacts to also 'rise' to the surface.

Even with optimal surface conditions the efficacy of fieldwalking varies according longterm land use, topography, weather conditions, the skill and experience of the fieldwalkers, and other factors. Intensive arable agriculture on hilltops will first expose and then pulverize artifacts such as pottery and even chipped stone (typically flint, chert or obsidian) flakes. Conversely, the plateau and upper scarp or valley side soils will move down slope, forming a deep seal over low-lying archaeological deposits, rendering them inaccessible to surface survey. Even artifacts on the surface and with relatively high visibility (i.e., little obscuring vegetation), however, are not consistently detected by surveyors. Consequently, it is unrealistic to expect 100% recovery of artifacts or even sites. We can evaluate surveyors effectiveness at detecting artifacts with "Sweep width," which is the theoretical width of a transect in which the number of artifacts detected outside the sweep is identical to the number missed within the sweep. The poorer the visibility, the poorer the contrast between the artifact "targets" and their surroundings, or the poorer the surveyor's skill or attention, the narrower the sweep width will be.

Modern technology such as GPS has made survey recording much easier, as positions of artifacts or artifact clusters ("sites") can be taken well within the limits of accuracy and precision necessary for survey work. Recording the position and attributes of archaeological features has been expedited by customizable portable computing interfaces or mobile Geographical Information Systems (GIS). Databases containing existing regional archaeological data as well as other landscape GIS layers such as soils, vegetation, modern features, and development plans can be loaded on a mobile GIS for referencing, for sampling purposes, and for groundtruth updating directly in the field, resulting a more informed archaeological survey process.

Visible Above-ground Structures

Fieldwalking picks up artefact scatters in ploughed fields. In heavily wooded areas such as Scandinavia, fieldwalking is not a central surveying method. Here artefacts and structures are hidden by humus and fallen leaves, and are therefore virtually invisible even at short distances. Instead archaeological surveyors concentrate on the region's abundant above-ground structures such as burial cairns, collapsed field walls and rock art panels, looking for unnatural changes in the vegetation and landscape to decide what may be hidden under the vegetation, or more rarely surveying by subsurface testing (SST). SSTs can consist of a series of shovel-test pits dug down below this humus layer or, where substantial later sediments may cover archaeological materials, series of auger or core holes. Because SSTs are much more costly than fieldwalking, surveys by SST usually have very low intensity. The various Scandinavian sites and monuments registers mainly list above-ground monuments, not ploughed-out pottery scatters.

Narrowing it Down

Because of the high costs involved in some kinds of surveys, it is often helpful to use "predictive modelling" to narrow down the search for archaeological materials. This is particularly important for purposive surveys, but can also be used to guide sampling surveys by eliminating the need to survey areas where, for geological or other reasons, we can reasonably expect all ancient traces to be destroyed (e.g., by erosion) or far too deeply buried (e.g., by alluvium) to be detectable. Modern predictive models in archaeology employ Geographic Information Systems (GIS).

Geophysical Survey

Geophysical survey is used for subsurface mapping of archaeological sites. In recent years, there have been great advances in this field, and it is becoming an increasingly useful and cost-effective tool in archaeology. Geophysical instruments can detect buried archaeological features when their electrical or magnetic properties contrast measurably with their surroundings. In some cases, individual artifacts, especially metal, may be detected as well. Readings taken in a systematic pattern become a dataset that can be rendered as image maps for interpretation. Survey results can be used to guide excavation and to give archaeologists insight into the patterning of non-excavated parts of the site. Unlike other archaeological methods, geophysical survey is not invasive or destructive. For this reason, it is often used where preservation (rather than excavation) is the goal for project preservation and compliance with applicable laws.

The geophysical methods most commonly applied to archaeology are magnetometers, electrical resistance metres, ground-penetrating radar (GPR) and electromagnetic (EM) conductivity. These methods provide excellent resolution of many types of archaeological features, and are capable of high sample density surveys of very large areas and of operating under a wide range of conditions. While common metal detectors are geophysical sensors, they are not capable of generating high-resolution imagery. Other established and emerging technologies are also finding use in archaeological applications.

Although geophysical survey has been used in the past with intermittent success, good results are very likely when it is applied appropriately. It is most useful when it is used in a well-integrated research design where interpretations can be tested and refined. Interpretation requires a knowledge both of the archaeological record, and of the way it is expressed geophysically. Appropriate instrumentation, field survey design, and data processing are essential for success, and must be adapted to the unique geology and archaeological record of each site. In the field, control of data quality and spatial accuracy are critical to a successful mission completion.

Analysis

The most important part of the survey is the analysis. The types of questions typically asked of survey data include: what is the evidence for first occupation of an area; when was this area occupied; how are sites distributed; where are sites located; what evidence is there for a settlement hierarchy; what sites are contemporary with each other; how has the modern landscape interfered with the visibility of archaeological remains; what sorts of activities can be recognized (e.g. dwellings, tombs, field systems); how many people lived in this area (at any given time); why did people choose to live in this area; how has the landscape changed over time; what changes in settlement patterns have there been?

At times, one part of the survey may not have yielded the evidence one wanted to find. For instance, very little may have been found during a field walk, but there are strong indications from geophysical survey and local stories that there is a building underneath a field. In such a case, the only way to decide if an excavation is worth the cost is to carefully analyze the evidence to determine which part to trust. On the one hand, the geophysics might just show an old and forgotten water-pipe, but it might also show the wall of just the building the archaeologists were looking for. The analysis therefore includes careful examination of all the evidence collected. A method often used to determine its value is to compare it to sites of the same period. As the number of well-documented surveys grow, this becomes a slightly easier task, as it is sometimes easier to compare two survey results than to compare a survey result with an excavated site.

Jacob's Staff

The term Jacob's staff, also cross-staff, a ballastella, a fore-staff, or a balestilha, is used to refer to several things. This can lead to considerable confusion unless one clarifies the purpose for the object so named. The two most frequent uses are:

- in astronomy and navigation for a simple device to measure angles, later replaced by the more precise sextants;
- insurveying for a vertical rod that penetrates the ground and supports a compass or other instrument.

Astronomy and Navigation

In navigation the instrument is also called a cross-staff and was used to determine angles, for instance the angle between the horizon and Polaris or the sun to determine a vessel's latitude, or the angle between the top and bottom of an object to determine the distance to said object if its height is known, or the height of the object if its distance is known, or the horizontal angle between two visible locations to determine one's point on a map.

The Jacob's staff, when used for astronomical observations, was also referred to as a radius astronomicus. With the demise of the cross-staff, in the modern era the name "Jacob's staff" is applied primarily to the device used to provide support for surveyor's instruments.

History

The origin of the name of the instrument is not certain. Some refer to the Biblical patriarch Jacob, specifically Gen 32:11. It may also take its name after its resemblance to Orion, referred to by the name of Jacob on some medieval star charts. Another possible source is the Pilgrim's staff, the symbol of St James (Jacobus in Latin). The name *cross staff* simply comes from its cruciform shape.

The original Jacob's staff was developed as a single pole device in the 14th century that was used in making astronomical measurements. It was first described by the Jewish mathematician Levi ben Gerson of Provence. However, its invention was likely due to Jacob ben Makir who also lived in Provence in the same period. Attributions to 15th century astronomer Georg Purbach are less likely correct, since Purbach was not born until 1423. Such attributions may refer to a different instrument with the same name. May states that its origins can be traced to the Chaldeans around 400 BC.

Although it has become quite accepted that Levi ben Gerson first described Jacob's staff, the Sinologist Joseph Needham theorizes that the Song DynastyChinese scientist Shen Kuo (1031–1095), in his *Dream Pool Essays* of 1088, described a Jacob's staff. Shen was an antiquarian interested in ancient objects; after he unearthed an ancient crossbow-like device from a home's garden in Jiangsu, he realized it had a sight with a graduated scale that could be used to measure the heights of distant mountains, likening it to how mathematicians measure heights by using right-angle triangles. He wrote that when one viewed the whole breadth of a mountain with it, the distance on the instrument was long; when viewing a small part of the mountainside, the distance was short; this, he wrote, was due to the cross piece that had to be pushed further away from the eye, while the graduation started from the further end. Needham does not mention any practical application of this observation.

During the Renaissance, the Dutch mathematician and surveyor Metius is known to have developed his own Jacob's staff. Gemma Frisius is also known to have made improvements to this instrument. Johannes Müller, called Regiomontanus, made the Jacob's staff in the 15th century to a popular instrument in geodesic and astronomical measurements.

Construction

In the original form of the cross-staff, the pole or main staff was marked with graduations for length. The cross-piece (*BC* in the drawing to the right), also called the *transom* or *transversal*, slides up and down on the main staff. On older instruments, the ends of the transom were cut straight across. Newer instruments had brass fittings on the ends with holes in the brass for observation. In marine archaeology, these fittings are often the only components of a cross-staff that survive. It was common to provide several transoms, each with a different range of angles it would measure. Three transoms were common. In later instruments, separate transoms were switched in favour of a single transom with pegs to indicate the ends. These pegs mounted in one of several pairs of holes symmetrically located on either side of the transom. This provided the same capability with fewer parts. The transom on Frisius' version had a sliding vane on the transom as an end point.

Usage

The navigator places one end of the main staff against his cheek just below his eye. He sights the horizon at the end of the lower part of the transom (or through the hole in the brass fitting) (*B*), adjusting the cross arm on the main arm until he or she can sight the sun at the other end of the transom (*C*). The altitude can then be determined by reading the position of the transom on the scale on the main staff. This value was converted to an angular measurement by looking up the value in a table.

Cross-staff for Navigation

The original version was not used at sea. Johannes Werner suggested the cross-staff be used at sea in 1514 and improved instruments were introduced for use in navigation. John Dee introduced it to England in the 1550s. In the improved versions, the rod was graduated directly in degrees. This variant of the instrument is not correctly termed a *Jacob's staff* but is a cross-staff. The cross-staff was difficult to use. In order to get consistent results, the observer had to position the end of the pole precisely against his cheek. He had to observe the horizon and a star in two different directions while not moving the instrument when he shifted his gaze from one to the other. In addition, observations of the sun required the navigator to look directly at the sun. This could be a painful exercise and made it difficult to obtain an accurate altitude for the sun. Mariners took to mounting smoked-glass to the ends of the transoms to reduce the glare of the sun.

As a navigational tool, this instrument was eventually replaced, first by the backstaff or quadrant, neither of which required the user to stare directly into the sun, and later by the octant and the sextant. Perhaps influenced by the backstaff, some navigators modified the cross-staff to operate more like the former. Vanes were added to the ends of the longest cross-piece and another to the end of the main staff. The instrument was reversed so that the shadow of the upper vane on the cross piece fell on the vane at the end of the staff. The navigator held the instrument so that he would view the horizon lined up with the lower vane and the vane at the end of the staff. By aligning the horizon with the shadow of the sun on the vane at the end of the staff, the elevation of the sun could be determined. This actually increased the accuracy of the instrument, as the navigator no longer had to position the end of the staff precisely on his cheek.

Another variant of the cross-staff was a spiegelboog, invented in 1660 by the Dutchman, Joost van Breen. Ultimately, the cross-staff could not compete with the backstaff in many countries. In terms of handling, the backstaff was found to be more easy to use. However, it has been proven by several authors that in terms of accuracy, the cross-staff was superior to the backstaff. Backstaves were no longer allowed on board Dutch East India Company vessels as per 1731, with octants not permitted until 1748.

Surveying

In surveying the *Jacob's staff*, contemporaneously referred to as a *jacob staff*, is a single straight rod or staff made of nonferrous material, pointed and metal-clad at the bottom for penetrating the ground. It also has a screw base and occasionally a ball joint on the mount, and is used for supporting a compass, transit, or other instrument. The term *cross-staff* may also have a different meaning in the history of surveying. While the astronomical cross-staff was used in surveying for measuring angles, two other devices referred to as a cross-staff were also employed.

1. Cross-head, cross-sight, surveyor's cross or cross - a drum or box shaped device mounted on a pole. It had two sets of mutually perpendicular sights. This device was used by surveyors to measure offsets. Sophisticated versions had a compass and spirit levels on the top. The French versions were frequently eight-sided rather than round.
2. Optical square - an improved version of the cross-head, the optical square used two mirrors at 45° to each other. This permitted the surveyor to see along both axes of the instrument at once.

Use of the Jacob's Staff as a Support

In the past, many surveyor's instruments were used on a Jacob's staff. These include:

- Cross-head, cross-sight, surveyor's cross or cross
- Graphometer
- Circumferentor
- Holland circle
- Miner's dial
- Optical square
- Surveyor's Sextant
- Surveyor's target

Some devices, such as the modern optical targets for laser-based surveying, are still in common use on a Jacob's staff.

High Water Mark

A high water mark is a point that represents the maximum rise of a body of water over land. Such a mark is often the result of a flood, although high water marks may reflect an all-time high, an annual high (i.e., highest level to which water rose that year), or the high point for some other division of time. Knowledge of the high water mark for an area is useful in managing the development of that area, particularly in making preparations for flood surges. High water marks from floods have been measured for planning purposes since at least as far back as the civilizations of ancient Egypt. It is a common practice to create a physical marker indicating one or more of the highest water marks for an area, usually with a line at the level to which the water rose, and a notation of the date on which this high water mark was set. This may be a free-standing flood level sign or other marker, or it may be affixed to a building or other structure that was standing at the time of the flood that set the mark. A high water mark is not necessarily an actual physical mark, although it is possible for water rising to a high point to leave a lasting physical impression, such as floodwater staining. A landscape marking left by the high water mark of ordinary tidal action may be called a strandline, and is typically composed of debris left by high tide. The area at the top of a beach where debris is deposited is an example of this phenomenon. Where there are tides, this line is formed by the highest position of the tide, and moves up and down the beach on a fortnightly cycle. The debris is chiefly composed of rottingseaweed, but can also include a large amount of litter, either from ships at sea or from sewage outflows.

Ecological Significance

The strandline is an important habitat for a variety of animals. In parts of the United Kingdom, Sandhoppers such as *Talitrus saltator*, and the seaweed fly*Coelopa frigida* are abundant in the rotting seaweed, and these invertebrates provide food for shore birds such as the rock pipit, turnstone and pied wagtail, and mammals such as brown hares, foxes, voles and mice.

Legal Significance

One kind of high water mark is the ordinary high water mark or average high water mark, which is the high water mark that can be expected to be produced by a body of water in non-flood conditions. The ordinary high water mark may have legal significance, often being used to demarcate property boundaries. The ordinary high water mark has also been used for other legal demarcations. For example, a 1651 analysis of laws passed by the English Parliament notes that for persons granted the title Admiral of the English Seas, "the Admirals power extended even to the high water mark, and into the main streams".

In the United States, the high water mark is also significant because the United States Constitution gives Congress the authority to legislate for waterways, and the high water mark is used to determine the geographic extent of that authority. Federal regulations (33 CFR 328.3(e)) define the "ordinary high water mark" (OHWM) as "that line on the shore established by the fluctuations of water and indicated by physical characteristics such as a clear, natural line impressed on the bank, shelving, changes in the character of soil, destruction of terrestrial vegetation, the presence of litter and debris, or other appropriate means that consider the characteristics of the surrounding areas. For the purposes of Section 404 of the Clean Water Act, the OHWM defines the lateral limits of Federal jurisdiction over non-tidal water bodies, in the absence of adjacent wetlands. For the purposes of Sections 9 and 10 of the Rivers and Harbors Act of 1899, the OHWM defines the lateral limits of Federal jurisdiction over traditional navigable waters of the U.S. The OHWM is used by the U.S. Army Corps of Engineers, the U.S. Environmental Protection Agency, and other Federal agencies to determine the geographical extent of their regulatory programmes. Likewise, many states use similar definitions of the OHWM for the purposes of their own regulatory programmes.

Metaphorical Usage

The term, "high water mark" has been used as a metaphor for the maximum extent of geographic control of an occupier of territory, and later for the maximum level of power, popularity, or some other characteristic, enjoyed by an entity. Some examples of metaphorical usage include:

- High-water mark of the Confederacy, the turning point of the Battle of Gettysburg.
    - o High Water Mark of the Rebellion Monument, a Gettysburg Battlefield memorial.
- High water marks, a hedge fund term used to determine fund performance fees.
- High-water mark (computer security), a computer security model wherein a document takes on the highest level of confidentiality allowed to the last person to access it.

Clairaut's Theorem

Clairaut's theorem, published in 1743 by Alexis Claude Clairaut in his *Théorie de la figure de la terre, tirée des principes de l'hydrostatique*, synthesized physical and geodetic evidence that the Earth is an oblate rotational ellipsoid. It is a general mathematical law applying to spheroids of revolution. It was initially used to relate the gravity at any point on the Earth's surface to the position of that point, allowing the ellipticity of the Earth to be calculated from measurements of gravity at different latitudes.

Formula

Clairaut's formula for the acceleration due to gravity g on the surface of a spheroid at latitude φ, was:

$$g = G\left[1 + \left(\frac{5}{2}m - f\right)\sin^2\phi\right],$$

where G is the value of the acceleration of gravity at the equator, m the ratio of the centrifugal force to gravity at the equator, and f the flattening of a meridian section of the earth, defined as:

$$f = \frac{a-b}{a},$$

(where a = semimajor axis, b=semiminor axis).

Clairaut derived the formula under the assumption that the body was composed of concentric coaxial spheroidal layers of constant density. This work was subsequently pursued by Laplace, who relaxed the initial assumption that surfaces of equal density were spheroids. Stokes showed in 1849 that the theorem applied to any law of density so long as the external surface is a spheroid of equilibrium. A history of the subject, and more detailed equations for g can be found in Khan.

The above expression for g has been supplanted by the Somigliana equation:

$$g = G\left[\frac{1 + k\sin^2\phi}{\sqrt{1 - e^2\sin^2\phi}}\right],$$

where, for the Earth, G =9.7803267714 ms^{-2}; k =0.00193185138639 ; e^2 =0.00669437999013.

Clairaut's Relation

A formal mathematical statement of the (unrelated) Clairaut's theorem is:

Let γ be a geodesic on a surface of revolution S, let ρ be the distance of a point of S from the axis of rotation, and let ψ be the angle between γ and the meridians of S. Then $\rho \sin \psi$ is constant along γ. Conversely, if $\rho \sin \psi$ is constant along some curve γ in the surface, and if no part of γ is part of some parallel of S, then γ is a geodesic.

Pressley (p. 185) explains this theorem as an expression of conservation of angular momentum about the axis of revolution when a particle slides along a geodesic under no forces other than those that keep it on the surface.

Geodesy

The spheroidal shape of the Earth is the result of the interplay between gravity and centrifugal force caused by the Earth's rotation about its axis. In his *Principia,* Newton proposed the equilibrium shape of a homogeneous rotating Earth was a rotational ellipsoid with a flattening f given by 1/230. As a result gravity increases from the equator to the poles. By applying Clairaut's theorem, Laplace was able to deduce from 15 gravity values that $f = 1/330$. A modern estimate is 1/298.25642.

Construction Surveying

Construction surveying (otherwise known as "lay-out" or "setting-out") is to stake out reference points and markers that will guide the construction of new structures such as roads or buildings. These markers are usually staked out according to a suitable coordinate system selected for the project.

History of Construction Surveying

- The nearly perfect squareness and north-south orientation of the Great Pyramid of Giza, built c. 2700 BC, affirm the Egyptians' command of surveying.
- A recent reassessment of Stonehenge (c.2500 BC) suggests that the monument was set out by prehistoric surveyors using peg and rope geometry.
- In the sixth century BC geometric based techniques were used to construct the tunnel of Eupalinos on the island of Samos.

Elements of the Construction Survey

- Survey existing conditions of the future work site, including topography, existing buildings and infrastructure, and underground infrastructure whenever possible (for example, measuring invert elevations and diameters of sewers at manholes);
- Stake out reference points and markers that will guide the construction of new structures
- Verify the location of structures during construction;
- Conduct an As-Built survey: a survey conducted at the end of the construction project to verify that the work authorized was completed to the specifications set on plans.

Coordinate Systems used in Construction

Land surveys and surveys of existing conditions are generally performed according to geodesic coordinates. However for the purposes of construction a more suitable coordinate system will often be used.

During construction surveying, the surveyor will often have to convert from geodesic coordinates to the coordinate system used for that project.

Chainage or Station

In the case of roads or other linear infrastructure, a *chainage* (derived from Gunter's Chain - 1 chain is equal to 66 feet or 100 links) will be established, often to correspond with the centre line of the road or pipeline. During construction, structures would then be located in terms of *chainage*, *offset* and *elevation*. *Offset* is said to be "left" or "right" relative to someone standing on the *chainage line*who is looking in the direction of increasing *chainage*. Plans would often show *plan* views (viewed from above), *profile* views (a "transparent" section view collapsing all section views of the road parallel to the *chainage*) or *cross-section* views (a "true" section view perpendicular to the *chainage*). In a *plan* view, *chainage* generally increases from left to right, or from the bottom to the top of the plan. *Profiles* are shown with the chainage increasing from left to right, and *cross-sections* are shown as if the viewer is looking in the direction of increasing *chainage* (so that the "left" *offset* is to the *left* and the "right" *offset* is to the *right*).

Building Grids

In the case of buildings, an arbitrary system of grids is often established so as to correspond to the rows of columns and the major load-bearing walls of the building. The grids may be identified alphabetically in one direction, and numerically in the other direction (as in a road map). The grids are usually but not necessarily perpendicular, and are often but not necessarily evenly spaced. Floors and basement levels are also numbered. Structures, equipment or architectural details may be located in reference to the floor and the nearest intersection of the arbitrary axes.

Other Coordinate Systems

In other types of construction projects, arbitrary "north-south" and "east-west" reference lines may be established, that do not necessarily correspond to true coordinates.

Equipment and Techniques Used in Construction Surveying

Surveying equipment, such as levels and theodolites, are used for accurate measurement of angular deviation, horizontal, vertical and slope distances. With computerisation, electronic distance measurement (EDM), total stations, GPS surveying and laser scanning have supplemented (and to a large extent supplanted) the traditional optical instruments.

The builder's level measures neither horizontal nor vertical angles. It simply combines a spirit level and telescope to allow the user to visually establish a line of sight along a level plane. When used together with a graduated staff it can be used to transfer elevations from one location to another. An alternative method to transfer elevation is to use water in a transparent hose as the level of the water in the hose at opposite ends will be at the same elevation.

Equipment and Techniques Used in Mining and Tunnelling

Total stations are the primary survey instrument used in mining surveying.

Underground Mining

A total station is used to record the absolute location of the tunnel walls (stopes), ceilings (backs), and floors as the drifts of an underground mine are driven. The recorded data is then downloaded into a CAD programme, and compared to the designed layout of the tunnel. The survey party installs control stations at regular intervals. These are small steel plugs installed in pairs in holes drilled into walls or the back. For wall stations, two plugs are installed in opposite walls, forming a line perpendicular to the drift. For back stations, two plugs are installed in the back, forming a line parallel to the drift.

A set of plugs can be used to locate the total station set up in a drift or tunnel by processing measurements to the plugs by intersection and resection.

Professional Status of Construction Surveyors

Building Surveying emerged in the 1970s as a profession in the United Kingdom by a group of technically minded General Practice Surveyors. Building Surveying is a recognized profession within Britain and Australia. In Australia in particular, due to risk mitigation/ limitation factors the employment of surveyors at all levels of the construction industry is widespread. There are still many countries where it is not widely recognized as a profession. The Services that Building Surveyors undertake are broad but include:

- Construction design and building works
- Project Management and monitoring
- CDM Co-ordinator under the Construction (Design & Management) Regulations 2007
- Property Legislation adviser
- Insurance assessment and claims assistance

- Defect investigation and maintenance adviser
- Building Surveys and measured surveys
- Handling Planning applications
- Building Inspection to ensure compliance with building regulations
- Undertaking pre-acquisition surveys
- Negotiating dilapidations claims

Building Surveyors also advise on many aspects of construction including:

- design
- maintenance
- repair
- refurbishment
- restoration

Clients of a building surveyor can be the public sector, Local Authorities, Government Departments as well as private sector organisations and work closely with architects, planners, homeowners and tenants groups. Building Surveyors may also be called to act as an expert witness. It is usual for building surveyors to undertake an accredited degree qualification before undertaking structured training to become a member of a professional organisation. For Chartered Building Surveyors, these courses are accredited by the Royal Institution of Chartered Surveyors. Other professional organisations that have building surveyor members include CIOB, ABE, HKIS and RICS.

With the enlargement of the European community, the profession of the Chartered Building Surveyor is becoming more widely known in other European states, particularly France. Chartered Building Surveyors, where many English speaking people buy second homes.

Distinction From Land Surveyors

In the United States, Canada, the United Kingdom and most Commonwealth countries land surveying is considered to be a distinct profession. Land surveyors have their own professional associations and licencing requirements. The services of a licenced land surveyor are generally required for boundary surveys (to establish the boundaries of a parcel using its legal description) and subdivision plans (a plot or map based on a survey of a parcel of land, with boundary lines drawn inside the larger parcel to indicated the creation of new boundary lines and roads).

Orthometric Height

The orthometric height of a point is the distance H along a plumb line from the point to the geoid. Orthometric height is for practical purposes "height above sea level" but the current NAVD88 datum is tied to a defined elevation at one point rather than to any location's exact mean sea level. Orthometric heights are usually used in the US for engineering work, although dynamic height may be chosen for large-scale hydrological purposes. Heights for measured points are shown on National Geodetic Survey data sheets, data that was gathered over many decades by precise spirit levelling over thousands of miles.

Since gravity isn't constant over large areas the orthometric height of a level surface isn't constant, and NGS orthometric heights are corrected for that effect. For example, gravity is 0.1% stronger in the northern United States than in the southern, so a level surface that has an orthometric height of 1000 metres in Montana will be 1001 metres high in Texas.

Practical applications must use a model rather than measurements to calculate the change in gravitational potential versus depth in the earth, since the geoid is below most of the land surface (e.g., the Helmert Orthometric heights of NAVD88). GPS measurements give earth-centred coordinates, usually displayed as height above the reference ellipsoid, which cannot be related accurately to orthometric height above the geoid without accurate gravity data for that location. NGS is undertaking the GRAV-D ten-year programme to obtain such data. Alternatives to orthometric height include dynamic height and normal height.

Tape Measure

A tape measure or *measuring tape* is a flexible form of ruler. It consists of a ribbon of cloth, plastic, fibre glass, or metal strip with linear-measurement markings. It is a common measuring tool. Its design allows for a measure of great length to be easily carried in pocket or toolkit and permits one to measure around curves or corners. Today it is ubiquitous, even appearing in miniature form as a keychain fob, or novelty item. Surveyors use tape measures in lengths of over 100 m (300+ ft).

Uses

Tape measures that were intended for use in tailoring or dressmaking were made from flexible cloth or plastic. Today, measuring tapes made for sewing are made of fiberglass, which does not tear or stretch as easily. Measuring tapes designed for carpentry or construction often use a stiff, curved metallic ribbon that can remain stiff and straight when extended, but retracts into a coil for convenient storage. This type of tape measure will have a floating tang on the end to aid measuring. The tang will float a distance equal to its thickness, to provide both inside and outside measurements that are accurate. A tape measure of 25 or even 100 feet can wind into a relatively small container. The self-marking tape measure allows the user an accurate one hand measure.

History

The first record of the use of the tape measure was by the Romans using marked strips of leather. On 3 January 1922, Hiram Farrand received the patent he filed in 1919. Sometime between 1922 and December 1926, Farrand experimented with the help of The Brown Company in Berlin, New Hampshire. It is there Hiram and W.W. Brown began mass-producing the tape measure. Their product was later sold to Stanley Works.

Design

The design on which most modern spring tape measures are built was patented by a New Haven, Connecticut resident named Alvin J. Fellows on 14 July 1868. According to the text of his patent, Fellows' tape measure was an improvement on other versions previously designed. The spring tape measure has existed since Fellows' patent in 1868, but did not come into wide usage until the early 1900s, when it slowly began to supplant a common folding wooden design of carpenter's ruler.

With the mass production of the integrated circuit (IC) the tape measure has also entered into the digital age with the digital tape measure. Some incorporate a digital screen to give measurement readouts in multiple formats. There are also other styles of tape measures that have incorporated lasers and ultrasonic technology to measure the distance of an object with fairly reliable accuracy.

Bibliography

Abidin, H.Z.: *Computational and Geometrical Aspects of on-the-fly Ambiguity Resolution*, University of New Brunswick, Canada, 1993.

Barnes, Thomas G.: *Origin and Destiny of the Earth's Magnetic Field,* Creation Research Society, London, 1983.

Bhatta, B.: *Remote Sensing and GIS,* Oxford University Press, Delhi, 2008.

Brown B.: *Life and Death of Coral Reefs*, Chapman and Hall, New York, 1997.

Carey, S. Warren : *Theories of the Earth and Universe*, Stanford, Stanford Univ. Press, 1988.

Chen, H. S.: *Space Remote Sensing Systems: An Introduction*, Orlando, Academic Press, 1985.

Clarke, B.: *Aviator's Guide to GPS*, McGraw-Hill, NY, 1994.

Crumley, Carole L.: *Celtic Social Structure: The Generation of Archaeologically Testatale Hypotheses from Literary Evideaicc,* University of Michigan, USA, 1974.

David T.: *Land Use Planning and Remote Sensing*, Canada, Kluwer Academic Pub., 1985.

Edwards, Denis: *Earth Revealing Earth Healing: Ecology and Christian Theology,* Collegeville: Liturgical Press, 2001.

Engman, E. T., and R. J. Gurney: *Remote Sensing in Hydrology*, New York, Van Nostrand Reinhold, 1991.

Gerald F.: *Marshall Handbook of Optical and Laser Scanning*, Marcel Dekker, Inc., 2004,

Golledge, R. and R. Stimson: *Spatial Behavior: A Geographic Perspective.* New York: The Guilford Press, 1997.

Hartshorne, Richard: *Perspective on the Nature of Geography*, Chicago, Rand McNally, 1959.

Johnson, Cait : *Earth, Water, Fire and Air: Essential Ways of Connecting to Spirit*, Woodstock, VT: Skylight Paths, 2003.

Juppenlatz, Morris : *Geographic Information Systems and Remote Sensing*, McGraw Hill, New York, 1996.

Kenney, D.: *Resource Management at the Watershed Level.* Natural Resources Law Center. Boulder, 1997.

Kuhn, Thomas S. : *The Structure of Scientific Revolutions,* Chicago: University of Chicago Press, 1970.

Langran, G.: *Time in Geographic Information Systems.* Bristol: Taylor & Francis, 1992.

Mack, Pamela Etter: *Viewing the Earth: The Social Construction of the Landsat Satellite System*, Cambridge, MIT Press, 1900.

Measures, Raymond M.: *Laser Remote Sensing: Fundamentals and Applications*, Malabar, Fla., Krieger Pub. 1992.

Okoshi, T.: *Three-dimensional Imaging Techniques*, Academic Press Inc., New York, 1976.

Peuquet, D.: *Representations of Space and Time.* New York: Guilford Press, 2002.

Prasad, C.V.R.K.: *Elementary Exercises in Geology*, Universities Press, Delhi, 2005.

Przyrbyla, J.: *Practical Considerations for GIS.* Redlands, California: 2010 GeoDesign Summit, 06-08 January 2010.

Robinson, I. S.: *Satellite Oceanography: An Introduction for Oceanographers and Remote-sensing Scientists*, New York, Halsted Press, 1985.

Sharma, V K : *Origin and Development of Geology*, Vista International Pub, Delhi, 2008.

Sheets, Payson D. and Brian R. McKee: *Archaeology, Volcanism, and Remote Sensing in the Arenal Region, Costa Rica.* Austin: University of Texas Press, 1994.

Wolf, P. R.: *Elements of Photogrammetry,* New York, 1983.

Index

A

B

C

D

E

F

G

H

I

L

M

N

O

P

R

S

T

V

W

❑❑❑